HOMELAND SECURITY
OPERATIONAL ANALYSIS CENTER

Analyzing a More Resilient National Positioning, Navigation, and Timing Capability

RICHARD MASON, JAMES BONOMO, TIM CONLEY, RYAN CONSAUL, DAVID R. FRELINGER, DAVID A. GALVAN,
DAHLIA ANNE GOLDFELD, SCOTT A. GROSSMAN, BRIAN A. JACKSON, MICHAEL KENNEDY, VERNON R. KOYM,
JASON MASTBAUM, THAO LIZ NGUYEN, JENNY OBERHOLTZER, ELLEN M. PINT, PAROUSIA ROCKSTROH,
MELISSA SHOSTAK, KARLYN D. STANLEY, ANNE STICKELLS, MICHAEL J. D. VERMEER, STEPHEN M. WORMAN

Published in 2021

Preface

The 2017 National Defense Authorization Act (NDAA) mandated a study "to assess and identify the technology-neutral requirements to backup and complement the positioning, navigation, and timing [PNT] capabilities of the Global Positioning System [GPS] for national security and critical infrastructure."

Having accurate PNT is essential for critical infrastructures across the country, and currently GPS is the primary source of PNT information for many applications. However, GPS signals are susceptible to both unintentional and intentional disruption, thus leaving critical infrastructure vulnerable. Because of the essential need for precise positioning or timing in many critical-infrastructure sectors, the Homeland Security Operational Analysis Center (HSOAC) was tasked to provide the U.S. Department of Homeland Security (DHS) with a study of the costs and benefits of proposed new PNT systems for the various types of critical infrastructure.

This research was sponsored by the DHS National Protection and Programs Directorate, Office of Infrastructure Protection, and conducted within the Acquisition and Development Program of the HSOAC federally funded research and development center (FFRDC).

About the Homeland Security Operational Analysis Center

The Homeland Security Act of 2002 (Section 305 of Public Law 107-296, as codified at 6 U.S.C. § 185), authorizes the Secretary of Homeland Security, acting through the Under Secretary for Science and Technology, to establish one or more FFRDCs to provide independent analysis of homeland security issues. The RAND Corporation operates the HSOAC as an FFRDC for the DHS under contract HSHQDC-16-D-00007.

The HSOAC FFRDC provides the government with independent and objective analyses and advice in core areas important to the department in support of policy development, decisionmaking, alternative approaches, and new ideas on issues of significance. The HSOAC FFRDC also works with and supports other federal, state, local, tribal, and public- and private-sector organizations that make up the homeland security enterprise. The HSOAC FFRDC's research is undertaken by mutual consent with DHS and is organized as a set of discrete tasks. This report presents the results of research and analysis conducted under HSHQDC-17-J-00441, Analysis of Position Navigation Timing (PNT) Alternatives.

The results presented in this report do not necessarily reflect official DHS opinion or policy.

For more information on HSOAC, see www.rand.org/hsoac. For more information on this publication, see www.rand.org/t/RR2970.

Contents

Figures

Tables

Summary

The modern technological economy relies on the ability to locate specific points on the earth (positioning), accurately develop and execute routing to move people and goods from place to place (navigation), and synchronize events and actions across wide locations (timing). These needs have existed since early in human history, and devising better ways to meet them has itself been a focus of technological development. **Although modern technologies for positioning, navigation, and timing (PNT) have made these functions easier and faster, and although the democratization of PNT tools has been the basis for innovations that have shaped modern society, previous methods of meeting these needs still exist.** This reality has resulted in the emergence of a modern PNT ecosystem in which PNT is supported not only by such highly advanced technologies as satellite navigation systems and atomic clocks but also by foundational techniques and tools used for decades—or even centuries—to navigate on the high seas or locate points on the earth.

In considering the modern PNT ecosystem in the United States, focus often shifts rapidly to the Global Positioning System (GPS) constellation of satellites. GPS is a comprehensive source of PNT, providing signals that not only can be used to locate the precise position and altitude of an individual receiver on the planet but also can be used as a distributed source of timing data transmitted the world over. GPS is provided free of charge by the U.S. government and is integrated into many technologies and uses in the economy, from legacy critical infrastructure systems to the emerging sharing economy. **GPS is synonymous with PNT in many people's minds, but it is far from the only element of the national PNT ecosystem.** Global navigation satellite systems (GNSS) provided by other nations supplement the PNT signals coming from GPS. Precision clocks have decreased significantly in cost, making it practical for firms that rely on exact timing—such as financial trading firms, air traffic control, and communications firms—to affordably integrate clocks into their infrastructures. Other technologies for positioning and navigation signals are commercially available for implementation in wide areas, smaller geographic areas, and even in individual commercial facilities. Although some of these options are based on positioning or timing signals of their own (either transmitted over the air or by other means, such as fiber-optic or other communications networks), some take advantage of signals that are already available, including signals from cellular towers or even Wi-Fi access points. It is important to note that these complements have been enabled by GPS: Because of the capability to precisely locate terrestrial emitters (such as Wi-Fi nodes or cellular sites) using GPS, it has become straightforward to use their signals for positioning either alongside or as a substitute for GPS.

Finally, the ready availability of positioning information means that maps that are "on the shelf" today—regardless of whether that shelf is physical or electronic—are much more accurate and precise than ever before. It also means that those maps would be available for use even if real-time access to GPS (or other positioning sources) were disrupted. GPS, in conjunction with such tools as geographic information systems, has created many accurate geospatial data sets that preserve, at least for a time, the societal value of national PNT for use in commercial and civil applications. GPS was integral to the creation and benefits of this data legacy and is important in maintaining and extending it, but the daily activities of end users would retain many of the benefits should the GPS system not be usable for a period.

However, because GPS has had a keystone role in national PNT for years, **a question has arisen about whether its importance justifies national investment in backup capabilities, given the potential threats to its functioning.** Most recently, Congress, in the 2017 National Defense Authorization Act (NDAA), directed the relevant agencies to "jointly conduct a study to assess and identify the technology-neutral requirements to backup and complement the positioning, navigation, and timing capabilities of the Global Positioning System for national security and critical infrastructure."

This report presents the results of a study carried out in partial response to that direction. Conducted for the Department of Homeland Security (DHS), this report has a scope that is restricted to domestic, nonmilitary uses of GPS. To reflect the still quite large complexity of this technical and policy problem, we examined the potential for loss or degradation in GPS functioning by referring to explicit, realistic threat scenarios. Necessarily, this entailed the potential for events and adversaries to affect both GPS and other elements of the PNT ecosystem. Where other components of the ecosystem are already in place, we examined how they provide robustness and resilience to national PNT even in the absence of an explicit GPS backup. The presence of such elements of the broader PNT ecosystem reduces the potential costs of disruptions to GPS and, consequently, reduces the potential benefits of additional investments in GPS specific backups or investment in additional PNT robustness or resilience.

Threats to GPS

There are documented threats to the functioning of the GPS satellite system. In considering threats that could disrupt the functioning of the GPS satellite system, we examined publicly available information on both large-scale threats (e.g., a nuclear exchange between states affecting space systems, extreme space weather) and smaller-scale threats (e.g., the jamming or spoofing of GPS signals in local areas or cities). Confirmed cases of events across this spectrum exist, including solar storms, technical problems that have affected the GPS constellation, intentional spoofing of a vehicle's GPS systems to change the vehicle's movement, and jamming incidents of varying scope and duration.

According to previous analyses (including the DHS National Risk Estimate focused on GPS), the incidence rates of these threats range from very rare to much more common, although there are insufficient data available to make defensible probability estimates. Estimates of consequence similarly differ, with most threats likely disrupting or corrupting GPS signals for periods ranging from hours to days. A small number of larger-scale threats (e.g., nuclear war directly targeting space systems, large-scale cyber intrusion and sabotage) could affect GPS functioning for longer periods. Such scenarios could result in the loss of GPS for months or even years if many new satellites needed to be produced, in spite of the fact that national security demands would likely drive efforts to reconstitute capability as quickly as possible by building and launching replacement satellites. Aside from such low-probability scenarios in which the loss of PNT might not be the most immediate national concern, virtually all other disruption or loss scenarios involve effects lasting for days.

Across the different types of threats, the potential for other parts of the PNT ecosystem to be affected varies, meaning that both existing capabilities and potential backups or additions could be vulnerable. For threat scenarios involving strategic adversaries, attacks on GPS might be accompanied by attacks on other components of the PNT ecosystem—and the more concentrated or physically localized those components are, the greater their vulnerability to small, focused attacks. Other components of the ecosystem that are more distributed (e.g., maintenance of backup clocks or use of distributed signals, such as cellular or Wi-Fi, for positioning) appear quite robust to many types of threats to GPS, although they have a different set of potential threats to their functionality, such as insider threats at less-secure organizations or simply damage from severe weather.

Assessing the Costs Associated with GPS-Focused Threats

Although GPS is an important component of the national PNT ecosystem, it is far from the only source of capability for PNT. In the *National PNT Architecture Study* released by the National Security Space Office in 2008, the "as is" PNT architecture (as it existed at that time) was already envisioned as an ecosystem of different satellite systems; terrestrial time transfer systems; star-tracking systems; distributed clocks; deployed beacon systems; and lower-technology terrestrial tools, such as inertial navigation systems (INS) and compasses (National Security Space Office, 2008).

Today, more than a decade later, the technologies available and deployed to industries that rely on PNT to differing extents are even broader, including, for example, Galileo (the European GNSS), other satellite systems, and atomic clocks integrated into a chip (Microsemi, 2018).

Many of these alternative and complementary PNT capabilities are already implemented broadly, and some additional technologies are being implemented for public safety or other purposes. Markets have provided a strong incentive to acquire and maintain backup systems for financial traders, who need access to precise timing information to synchronize trades. In transportation systems, such as aviation where there are established risk-management practices suitable for life safety critical applications, both robustness and resilience to technical failures are prioritized. As a result, the diversity that exists in the current PNT ecosystem strengthens the country's and economy's resilience to disruptions in GPS, and the availability of these complements must be reflected in estimates of the cost of GPS outage or corruption. In some areas and for some functions, such as timing and coarse positioning in urban areas, the existing ecosystem appears to nearly fully address the risks associated with GPS disruption or corruption.

In addition, "fallback" technologies—for example, navigation by traditional visual or manual course plotting, positioning using reference points—increase the robustness of PNT nationally. In interviews with individuals in the transportation industry, we heard the argument that such techniques are routinely maintained, exercised, and needed to deal with technical failures, whether from natural or intentional disruptions to GPS, from computer system failures, or from communications breakdowns. Although many individuals use GPS every day for turn-by-turn driving directions, analyses show that the repetitiveness of most such trips—from home to work, school, or usual shopping locations—means that most people's lives would continue with limited interruption if PNT were disrupted. Even for trips that are nonroutine, tools storing previously collected geospatial information (either on board or in the cloud) can help with navigation and positioning in many circumstances. **Using such fallbacks is one way that dynamic human adaptation can reduce the costs of incidents affecting GPS, especially in the short and medium terms. If GPS disruption made it difficult to perform a task in an automated way, manual strategies might be available, even if the price of using these manual strategies is a reduction in the efficiency or effectiveness of the task.** Prospects are further brightened when we look to a variety of emerging new technologies. These technologies include improved radio-navigation features in next-generation smartphones, computer vision technologies that next-generation vehicles will use to recognize landmarks and refine their positions, and integrated timing and navigation systems making their way through military development channels.

In the past, analyses of the potential cost of loss or disruption to GPS applied **nationwide, sector-by-sector, microeconomic approaches, seeking to calculate**

the overall benefit to the economy from GPS availability. These analyses must be interpreted carefully, because the "benefit" from GPS can be defined or estimated in different ways, and the damages from a GPS outage may not correspond simply to lost benefits. Additionally, different analyses have focused on different sectors. Nevertheless, we believe that these analyses could be selectively used to inform this work in certain sectors. We used this earlier work, as well as more recent estimates from a team from RTI International (O'Connor et al., 2019), looking at estimates of GPS benefits in individual industries while also reflecting the different disruptions that could occur in different threat scenarios and the sensitivity of different industries to GPS loss or corruption.

To go beyond these existing analyses, we also conducted regionally based, sector-by-sector, microeconomic analyses that examined the ability of attackers to jam or spoof GPS signals, given the real-world constraints of geography and the built environment (as in cities). By discussing where jammers of different levels of power and reach would have to be placed in some cities—and therefore the number of jammers that would be needed to have more than a very localized effect—we also sought to reflect the potential for existing complements (e.g., backup clocks in the financial sector) to hedge against these potential threat scenarios.

Neither our regional nor nationwide analysis has attempted to estimate any new economic growth that might be spurred by the provision of a new service, presumably offering some improvement beyond GPS. We analyzed only the potential losses from existing economic activities, in keeping with the relatively short time horizon of our analyses.

In particular, we found that the threat from spoofing GPS signals should not influence a decision about any new PNT systems. Given the rich PNT ecosystem that has developed in the past decades, there are already many available ways to readily detect spoofing in any sensitive applications through integrity monitoring. Although such integrity monitoring may not, in fact, always be implemented, that is a problem in system design or regulation and is not due to the absence of an alternative signal for PNT.

Our conclusions from these various sets of analyses were largely consistent: When estimates of the cost of GPS disruption or loss include realistic adaptation options and existing complementary technologies, the estimates are surprisingly low.[1] Even for scenarios that would affect the entire nation, simplistic estimates in the billions (ignoring existing backups and adaptation) drop by orders of magnitude— into the tens to hundreds of millions of dollars per day—when one makes reasonable assumptions that include existing complementary PNT and backup approaches. Smaller-scale jamming or spoofing scenarios affecting even large fractions of important

---

[1] Some notable differences with two sectors in the RTI International analysis by O'Connor et al. (2019) are discussed in Chapter Four.

cities, which could be within the capacity of either state or nonstate adversaries, produce costs in the millions per day.[2] Further, reasonable assumptions about law enforcement's ability to identify and apprehend the perpetrators of such incidents, potentially supplemented with additional technology, suggest that maintaining disruptions of any significance over more than a few days would be difficult.

Thinking About Additions to the PNT Ecosystem in Light of the Potential Benefits

Because of the importance of PNT in the modern economy, a wide range of technologies has been proposed that could either supplement or—as it has been framed in national debate—back up GPS. Some of these systems are intended to cover wide areas. Enhanced Long-Range Navigation (eLoran) is one such system that has been proposed to explicitly back up GPS in response to the threats previously described. As a terrestrial system, it avoids some of the space-based vulnerabilities associated with GPS and other constellations, but it gains other vulnerabilities. There are also a variety of other technologies that could supplement the ecosystem in different ways, for example, by providing local alternative capabilities for PNT or performing other functions, such as monitoring GPS signals to detect disruption (and therefore defeat attempts at spoofing by warning users) or using sensors to detect the transmission of jamming signals in order to rapidly halt such incidents and enable apprehension of the perpetrators. Many of these alternative PNT technologies or supplemental systems already exist but have not been widely implemented. Others are in earlier stages of development.

No single system is a perfect backup for GPS. Some systems, such as the European Union's Galileo, are satellite systems very similar to GPS and could be a good substitute under certain circumstances (e.g., if GPS alone were spoofed or disrupted by a cyberattack). However, under other circumstances, such as a solar storm, these similar satellite systems would also be unavailable, as they share that vulnerability with GPS.

Terrestrial systems are dissimilar to GPS and have different weaknesses, but no single proposed terrestrial system matches GPS on area coverage; on position accuracy; and likely on ubiquitous adoption of low-cost user equipment in the presence of continual, free GPS signals. Any single proposed backup, therefore, will mitigate a GPS outage for only *some* users.

Because different critical infrastructure sectors and industries have different applications for PNT data, the level of accuracy and precision that is useful to them

---

[2]   This includes potential increased loss of life due to delays in emergency services; one statistical life values at $9.8 million in accordance with government guidelines. See Chapter Four for more detail.

differs. As a result, some of the potential alternatives or complements might not meet all industries' needs and, therefore, might be more or less attractive as additions to the national PNT ecosystem.

Detailed cost estimates for potential alternative PNT systems are proprietary to various firms and are therefore included in Appendix G (available separately), but we can independently estimate the approximate costs for these systems and compare them with the benefits.

The conclusion from our economic analysis is clear: Given realistic estimates of the costs of GPS disruption, the bar for extensive government investment in a backup to GPS (that does not serve some other purpose) is very high and therefore difficult to justify.

Most plausible disruptions of GPS will be limited either in space or time. Notably, of eight risk scenarios considered in the 2011 *National Risk Estimate* for GPS, only one scenario (geomagnetic storm) implied economy-wide disruption, and this would persist for only a few days.[3] If the costs of a national GPS disruption are at most in the tens to hundreds of millions per day—and if any single backup system would mitigate only a portion of those costs—then the risk does not justify more than modest government investment in any single backup system. An *extended* national outage would occur only under such low-probability scenarios as a limited space war where GPS satellites were directly targeted and destroyed, a case we find to be implausible.

For more localized threats, such as jammers intentionally used to disrupt GPS in a city, we found that more than a hundred such local attacks per decade would be needed to match the costs of the cheaper alternative PNT systems covering even a limited set of larger cities. Given the complete absence of such attacks to date, the threat of such attacks does not come even close to justifying investments in the backups or alternatives we examined.

Moreover, the federal government is already involved in one public-private partnership that will provide a backup to GPS for many users in important urban areas, albeit for other reasons. This development, which is part of the FirstNet partnership, is discussed in more detail in Chapters Five and Six. Although this backup is yet to be deployed, virtually all economic activity in the urban areas eventually covered by this system will already have an alternative signal to mitigate any GPS disruption, further lowering the benefit of another system.

Other possible federal efforts to protect the GPS signal could be based on a different cost model. For example, capabilities that pair GPS integrity monitoring and jamming detection would facilitate the detection and apprehension of individuals who could be fined as a result, partially defraying the cost of the system. For many perpetrators, such fines could be a strong deterrent for the most common (though smallest-scale) type of incident of concern. Although the scale of such an effort—and its associ-

---

[3] This is discussed in more depth in Chapter Three and, for the GPS satellites themselves, in Appendix F.

ated costs and benefits—would likely be quite small, it provides an example of such a change in the cost model.

Modest investments by the government in threat detection could also reinforce private incentives to maintain a robust PNT ecosystem. The private incentives would be calibrated directly by the private applications and uses of PNT data. The evolution of the ecosystem demonstrates that, where private entities rely on PNT and also have considerable value at risk, they will prudently invest in backups and complements to reduce risk on their own. Government investment in integrity monitoring and notification of PNT functioning would reinforce that incentive and provide such actors with the data they need to most effectively protect their own interests, given both natural and adversary risks to PNT. And they would not require any larger federal role.

Lessons on Maintaining a Robust and Resilient PNT Infrastructure

Although the utility of PNT and its integration into many economic and individual activities means that its disruption would be broadly felt, economic analysis that incorporates reasonable assumptions about the behavior of individuals and organizations suggests that the costs of disruption, while real, would not be as high as sometimes assumed. **As a result, government investment in a "GPS backup" appears unwarranted at this time. New PNT systems could be developed for the complementary benefits they bring while GPS is operating, but *not* primarily as an additional backup for GPS outages.**

Our approach to that analysis—including reasonable threat scenarios, considerations of attacker behavior, and a comprehensive examination of other technologies and adaptations—leads to several other overall conclusions for maintaining robust and resilient PNT for the United States.

First, having "time-proven, robust fallbacks" available is highly desirable. Maintaining those capabilities while seeking the efficiency gains of modern PNT should be a priority. For example, the fact that pilots—whether air or maritime—are not wholly dependent on GPS to operate their vehicles is a good thing. Public policy should not incentivize the shedding of such capabilities and expertise based on the unrealistic assumption that modern PNT has made it obsolete or unnecessary.

Second, dispersal and diversity of capabilities in the national PNT ecosystem is a strength, not a weakness. If an analysis invokes an adversary's ability to target GPS to justify investment in defenses or backups to hedge against such an attack, consistency requires inclusion of the potential for that same adversary to target the backup as well. Terrestrial backup systems are less exposed to space-based threats of concern for GPS, but a centralized system would be even more vulnerable to attack than GPS and would also be exposed to terrestrial natural risks, as from weather, that space-based systems avoid. Dispersed backup clocks or positioning and navigation that rely on such technologies as cell tower or Wi-Fi signals would localize the effects of such attack simply by virtue of being dispersed.

Third, considering both current and potential future systems, prudent system design necessitates not creating dependencies that increase the risk associated with GPS loss. For example, transportation systems—whether current ships or future autonomous vehicles—are only readily vulnerable to hijacking through GPS spoofing if they are designed to rely on that single source of data for navigation. For such designs, data validation and integrity are essential. Comparison of data from different sensors or sources is desirable, and such sources are already available: from the different frequency signals now used by GPS itself; from the signals of the other GNSS constellations and potentially Satellite Time and Location (STL), which uses the Iridium constellation; from the FirstNet system, soon to be available in urban areas; and, in some cases, from known landmarks. Engineering design that takes these potential threats into account is the needed step, independent of any additional signals or backups to the national PNT ecosystem.

Acknowledgments

We thank Jim Platt, Katherine Ledesma, Michael Striffolino, and Jay Robinson of the U.S. Department of Homeland Security (DHS) for supporting and guiding this research. Keith Conner and Ernest Wong of the DHS Science and Technology Directorate and Karen Van Dyke of the Department of Transportation provided valuable assistance.

Outside the federal government, several individuals generously answered our questions and explained technical aspects of some systems. In particular, we thank Ganesh Pattabiraman, Chris Gates, Subbu Meiyappan, and David Knutson of NextNav; Bill Goodman and Daniel Fortes of Ericsson; Dana Goward of the Resilient Navigation and Timing Foundation; Chris Kreger of RF Specialties Group; Dan Lutter of Seven Solutions; Andy Proctor of the United Kingdom Space Agency; Kiyeol Seo of the Korea Research Institute of Ships and Ocean Engineering; and Charles Schue of UrsaNav. Additional thanks are due to Stacey Worman for her insights into the difficulties associated with cost estimates related to space weather phenomena.

Within the Homeland Security Operational Analysis Center, Philip Antón, Richard Silberglitt, and John Parmentola provided helpful reviews. Our former colleague Doug Shontz also made contributions to this report.

Abbreviations

AIS	Automatic Identification System
ASF	Additional Secondary Factor (radio delay due to land terrain)
ATIS	Alliance for Telecommunications Industry Solutions
BEA	Bureau of Economic Analysis
CDMA	Code-Division Multiple Access
CME	coronal mass ejection
CONUS	continental United States
DFMC	dual-frequency multiconstellation
DGNSS	Differential Global Navigation Satellite Systems
DHS	U.S. Department of Homeland Security
DoD	U.S. Department of Defense
DoT	U.S. Department of Transportation
eLoran	Enhanced Long-Range Navigation
EMI	electromagnetic interference
EU	European Union
E911	enhanced 911
FAA	Federal Aviation Administration
FCC	Federal Communications Commission
FDD	Frequency-Division Duplexing
FFRDC	federally funded research and development center
GAO	U.S. Government Accountability Office
GDP	gross domestic product
GLONASS	Globalnaya Navigatsionnaya Sputnikovaya Sistema
GNSS	global navigation satellite systems
GPS	Global Positioning System
HSOAC	Homeland Security Operational Analysis Center

IMU	inertial measurement unit
INS	inertial navigation system
ISM	Industry Scientific and Medical (the 2.4-2.4835 GHz band)
Lidar	light detection and ranging
Loran	Long-Range Navigation
LTE	Long-Term Evolution
LTE-FDD	Long-Term Evolution–Frequency-Division Duplexing
MBS	Metropolitan Beacon System
MEO	medium earth orbit
MF	medium frequency (the 300 kHz–3 MHz band)
MiFID II	Markets in Financial Instruments Directive II
MSA	Metropolitan Statistical Area
NAICS	North American Industry Classification System
NDAA	National Defense Authorization Act
NIST	National Institute of Standards and Technology
NTP	Network Time Protocol
OMB	Office of Management and Budget
PF	primary factor (propagation delay due to atmosphere)
PFI	private finance initiative
PNT	position, navigation, and timing
PPP	public-private partnership
PSC	Public Sector Comparator
PTC	Positive Train Control
PTP	Precision Time Protocol
QZSS	Quasi-Zenith Satellite System
RF	radio frequency
RFI	request for information
RMS	root mean square
SDR	software-defined radio
SEP	solar energetic particle
SF	secondary factor (propagation delay due to the earth's surface)
SoOP	signal of opportunity
STL	Satellite Time and Location
TDD	Time-Division Duplexing

TID	total ionizing dose
USCG	U.S. Coast Guard
USDA	U.S. Department of Agriculture
USNO	U.S. Naval Observatory
UTC	Universal Time Coordinated
VfM	value for money
WLAN	wireless local area network
WWVB	a radio station call sign

CHAPTER ONE

Introduction

The modern technological economy relies on positioning, navigation, and timing (PNT)—specifically, the ability to locate specific points on the earth (positioning, P), the ability to accurately route people and goods from place to place (navigation, N), and the ability to synchronize events and actions in disparate locations (timing, T). Government roles and responsibilities—notably, national defense and the ability to effectively carry out military operations—similarly rely on these capabilities for logistics, synchronizing complex operations, and delivering forces or striking targets precisely.

The importance of PNT to both homeland security and military operations has naturally raised questions about the vulnerability of the United States to threats to the systems that provide PNT—most noticeably, the Global Positioning System (GPS).

Objective

This report responds to congressional direction contained in the National Defense Authorization Act (NDAA) of 2017 that directed the relevant agencies to "jointly conduct a study to assess and identify the technology-neutral requirements to back up and complement the positioning, navigation, and timing [PNT] capabilities of the Global Positioning System [GPS] for national security and critical infrastructure." Based on that guidance, the U.S. Department of Homeland Security (DHS) asked RAND's Homeland Security Operational Analysis Center (HSOAC) to conduct a study that responds to DHS's part of the NDAA mandate—identifying backup and complementary PNT capabilities to address the nonmilitary, homeland security needs, as opposed to identifying backup and complementary PNT capabilities to address the needs of the Department of Defense (DoD). In considering domestic transportation uses of PNT, our work was also supported by the Department of Transportation (DoT). Finally, and also in response to the congressional direction, we examined appropriate technology options and the viability of either a public-private partnership or service-level agreements to provide PNT capabilities.

Background

To provide some context for this objective, we provide some background on PNT systems and on past considerations about providing backups to them.

The Technical Ecosystem: Providing and Utilizing PNT in the United States

PNT did not start with GPS, but PNT has a very long history. Today, in part because of that history, PNT in the United States has both diverse sources and diverse uses.

Diverse PNT Sources

GPS is the most familiar system providing PNT. GPS is a satellite-based radionavigation system that relies on highly accurate, synchronized atomic clocks on each satellite that broadcast the position of the satellite and the time. By combining signals from multiple GPS satellites, a receiver can both precisely determine its position and altitude and simultaneously determine a highly accurate time anywhere in the world.

GPS was developed and fielded by the military beginning in 1978 and reached its full capability in 1995 (Ochieng et al., 2001). Because the PTN capabilities in GPS have value beyond the military, GPS was designated a dual-use system for civilian applications in 1996; four years later, features of the system that had degraded the quality of positioning data provided to nonmilitary users was turned off. GPS receivers became commercially available in 1981 (Wallischeck, 2016), starting at costs in the hundreds of thousands of dollars, but the economic benefit became much greater after advances in technology reduced the cost of receivers by orders of magnitude. GPS capability has now been integrated into a myriad of commercial applications and into virtually every smartphone now produced, thus putting GPS "in the pocket" of a large percentage of the individuals in the country. This transition can be traced back to National GPS Policy documents released in 1996, which state the intent for GPS Standard Positioning Services "[to be delivered] free of direct user fees for peaceful, civil, commercial, and scientific uses worldwide" (DoD, 2008). This intent drove adoption and applications with almost the strongest incentive possible: providing a valuable technical capability free of charge for all to use.

Although GPS is the crown jewel of U.S. domestic PNT capabilities, it does not make up the totality of PNT capabilities. As a result of cycles of technology development and parallel technology development in other nations, there are a diverse set of PNT sources relevant to different sectors of the economy. We summarize the main types of PNT sources in Figure 1.1.[1]

As shown at the top of the figure, three other satellite constellations similar to GPS are operational or expected to soon become so: Globalnaya Navigatsionnaya Sput-

---

[1]   For a more detailed look at the complex ecosystem of PNT sources, see the discussion in the 2008 *National PNT Architecture Study: Final Report* in its description of the then-architecture and its potential evolution out to 2025 (National Security Space Office, 2008).

Figure 1.1
Sketch of the National PNT Ecosystem

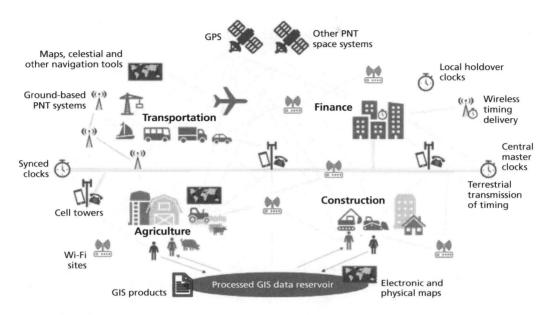

nikovaya Sistema or Global Navigation Satellite System (GLONASS), maintained by Russia; Galileo, maintained by the European Union (EU); and BeiDou, maintained by China. The United Kingdom is considering building yet another similar system given that it is leaving the EU and thus ceasing involvement with Galileo (Parker, Hollinger, and Barker, 2018). Given this, modern positioning receivers, including common cell phones, are increasingly designed to receive and use signals from multiple satellite constellations.

Additionally, multiple ground-based signals are used for timing and positioning, including signals from public and private Wi-Fi transmitters and cell phone towers, as well as dedicated terrestrially based position systems for wide areas[2] or for individual sites such as ports or warehouses.[3] Technologies even exist that use celestial position-

---

[2]   For example, Wide Area Multilateration (WAM) (FAA, 2016b).

[3]   Systems use a range of technologies designed to work cooperatively with the appropriate user equipment. Bluetooth and Wi-Fi (such as cell phones) are used in smaller closed areas for typical consumer-facing applications; light-based systems are used in radio frequency (RF)-sensitive areas; and RF-based systems like Locata are used over larger areas, such as ports and mines, where GPS reception might be unreliable or impossible. In addition, there are a host of systems used in hybrid configurations that measure state changes—such as accelerometers, compass headings, pressure altitude, and precision maps—to enable a rapid position fix for some applications. Vision-based systems are also entering the mix, such as on an autonomous system to observe landmarks and signage for navigation, in addition to their function for collision avoidance (Chipka and Campbell, 2018; Tariq et al., 2017).

ing—recreating in electronic form the same sorts of tasks relied on by ancient mariners crossing the sea.

For precision timing, the same technological forces that caused the price of GPS receivers to drop dramatically have also reduced the cost of alternative timing technologies. Although atomic clocks were at one time so expensive that only governments, major corporations, or research centers had them, the price of highly precise timepieces has fallen into the thousands of dollars. Atomic clocks are not nearly as inexpensive as GPS receivers, but this reduction in cost means that many economic actors, such as electric power companies or financial firms, can maintain their own precision timing infrastructure if timing data are critical to their operations and profitability.

In short, as shown in Figure 1.1, current technologies provide a range of overlapping PNT capabilities to government agencies, companies, nongovernmental organizations, and individuals.

Diverse PNT Uses

Beyond the diversity in sources of PNT, the uses of PNT vary considerably across the economy. Because of the ubiquity of PNT, we first think about the dynamic uses of these technologies for users in real-time motion; for example, mapping and navigating apps on smartphones help people navigate from place to place and provide support to the precision motion of machinery on farms, in ports, or at construction sites. However, there are other valuable uses of PNT that do not involve real-time motion, including providing positioning for surveyors, timing for tracking financial transactions, and timing for the synchronized functioning of such infrastructures as power and communications.

The benefits of GPS in the broader economic system accumulate over time and have created a huge set of location data. With widely proliferated mapping capability, more detailed survey information (e.g., of farms or other commercial properties) is more broadly available as are maps of urban areas and nautical or other charts. Even in the absence of real-time access to the GPS signal, such legacy data help enable navigation by other means and can help accomplish other tasks that rely on accurate maps and representations of land and built infrastructure.

The relative importance of the different uses of PNT varies across the economy. For example, for elements of the financial infrastructure, precision timing is highly critical for some functions but not for others. Thus, we must separately consider how different industries would respond to an outage of GPS.

Considering Backups to GPS

Because GPS has played a central role in integrating PNT into economic activity and daily life for decades, what might occur if something caused the system to become degraded or inoperable? In 2004, National Security Presidential Directive 39 recommended that relevant federal departments develop backup capabilities that could support national needs to address the potential scenario of "disruption of the Global Positioning System or other space-based positioning, navigation, and timing services." (National Coordination Office for Space-Based Positioning, Navigation, and Timing, 2004) In January 2007, when considering future policy with respect to other existing technical systems, DHS requested public comment on

> the need to continue to operate or invest in the North American [Long-Range Navigation] LORAN-C Radionavigation System beyond fiscal year 2007. Future investment decisions might include: decommissioning the LORAN-C system, maintaining the LORAN-C system as currently configured, or developing a fully deployed Enhanced LORAN (eLORAN) system. (Long Range Aids to Navigation [LORAN] Program; Office of Navigation and Spectrum Management, 2007)

The last option, developing an eLoran system, would provide terrestrial capabilities that could be viewed as a direct backup to GPS (Johnson et al., 2007).

Since 2007, numerous reports, including one by the National Security Space Office (2008), have made the point that PNT needs are being met through a variety of sources. The gaps in those systems led to the conclusion that perhaps an aggregated solution was better than several separate systems (GAO, 2013c, pp. 19–20). Additionally, federal policy with respect to GPS continued to reflect concern about the potential for GPS interference or failure, but varied considerably in focus. For example, the National Space Policy of the United States of America (The White House, 2010) required investment in capabilities to counter harmful interference to GPS. However, the policy directs the identification and implementation of backup systems or approaches only "as necessary and appropriate." Multiple queries from industry on both backups and requirements have oscillated between focusing on individual systems and focusing on a more balanced approach (Complementary Positioning, Navigation, and Timing Capability; Notice; Request for Public Comments, 2015; Positioning, Navigation, and Timing [PNT] Service for National Critical Infrastructure Resiliency, 2016). Reflecting a more technology-agnostic approach, the 2017 Federal Radionavigation Plan flagged potential economic and security benefits that might accrue by having a GPS backup, without mandating either a single- or multiple-system backup (DoD, DHS, and DoT, 2017).

Given the proliferation of PNT-related technologies outlined above, the description of the status quo in the 2008 PNT Architecture document remains apt. It states that the national PNT

> relies upon a large number of PNT enabling capabilities and infrastructures in an environment which includes spectrum, weather, fiscal and geo-political challenges . . . characterized by widespread use of GPS, and a large number of systems that augment GPS, where each augmentation is optimized for different user groups. (National Security Space Office, 2008)

But over that intervening decade, the technological context has continued to evolve. This evolving context suggests that the question of a backup to GPS must also evolve. Asking a narrowly defined question, such as whether a backup can ameliorate some specific hazard or threat to GPS, implicitly assumes some new system is needed to meet that threat. Instead, we believe it is better to approach the issue by considering both existing and future investment(s) across the PNT ecosystem in order to understand how the entire PNT system would respond to any degradation or loss of GPS. Within this evolving ecosystem—the diversity of which has only increased since the 2008 architecture effort—many factors drive the importance of PNT services and systems. In this complex system, the right public policy option can be difficult to guess, given that GPS is a valuable common utility and considering the need to ensure sustained support but hedge against unavailability.

Approach

Given this background and the objective of evaluating the need for a backup and/or complement to GPS, we developed an approach that considered each part of the issue in a logical fashion where each part of the analysis is needed as input to the next part of the analysis to evaluate the need for a backup and/or complement to GPS. Broadly, we decomposed the issue into four sequential parts: (1) identify risks with GPS and other PNT systems; (2) conduct an economic analysis of costs resulting from the varied threats to PNT; (3) identify alternative sources and technologies to increase national PNT resilience and robustness; and (4) conduct both quantitative and qualitative cost-benefit assessments, as appropriate, of alternative sources and technologies to increase national PNT resilience and robustness. Each of the four analysis parts focused on answering a series of questions, which are shown in Table 1.1.

Scoping

Overall, this report is designed to inform the DHS's response to the legislative requirement in the 2017 NDAA. As noted above, our analysis of alternative PNT systems does not consider the important national security uses of PNT because those are the responsibility of the DoD. This implies that we did not analyze the purely military uses of GPS or PNT nor any uses by the intelligence agencies. We did, however, include the use of PNT services for the defense industrial base within our general analyses of industrial dependencies on PNT. Additionally, our analysis was restricted to the uses of PNT within U.S. territory. This meant that we did not consider the utility of any alternative PNT systems for friendly or allied countries.

Table 1.1
Key Analysis Parts and Associated Questions

Key Analysis Parts	Associated Questions
1. Identify risks with GPS and other PNT systems	• What are the potential risks to GPS, from nation-states, sub-national groups, and natural events? • How large an area is affected from each, how severe is the disruption, and for how long would the disruptions last? • What can we say about the probability of such events? • What are the answers to these questions for alternative PNT systems?
2. Conduct an economic analysis of costs resulting from the varied threats to PNT	• How much damage is likely, by sector of the economy, aggregated for each risk consideration in the first part of the analysis? • What are estimates of the effects of existing backups and workarounds?
3. Identify alternative sources and technologies to increase national PNT resilience and robustness	• What are the mature technologies/systems that could supply alternative PNT signals? • How is each characterized by area covered, service provided (timing-only or full PNT), accuracy, applicability to different users (cell phone compatibility, requirement for fiber connection, etc.), vulnerability to various threats (importantly those sharing vulnerabilities with GPS, such as other global navigation satellite systems [GNSS]), and cost?
4. Conduct cost-benefit assessments of alternative sources and technologies to increase national PNT resilience and robustness	• How do the costs of available backups compare with the benefits those backups would provide? • In particular, which, if any, have costs less than or comparable to their benefits? • What are the opportunities for public-private partnerships in supplying the promising backups?

On the principle that user requirements should be determined prior to the search for solutions, the user requirements for PNT in various critical sectors were examined independently of our analysis by the Johns Hopkins Applied Physics Laboratory and by the Aerospace Corporation (Cavitt et al., 2018; Tralli et al., 2018a, 2018b), and we accepted those user needs as inputs to our study. Nonetheless, we did necessarily develop a deeper understanding of existing backups for the different sectors.

The technical options for alternative PNT systems were informed by the responses of industry to a request for information (RFI) issued by DHS. We helped draft that RFI and largely relied on the responses to it as well as to an earlier, similar RFI to the DoT. The responses included proprietary performance and cost data, which are only used here in Appendix G: Estimate of the Costs and Benefits of Alternative Proprietary PNT Systems.

Because of the particular interest in Congress, our analysis considered public-private partnerships to operate any backups or complements that seemed promising as well as service-level agreements if those would be appropriate. In general, we described the preferred management constructs for any promising systems.

Finally, we were instructed to use no classified information in our analysis.

Outline of This Report

The report is organized around the four sequenced analysis parts in Table 1.1 and answers the associated questions for each part. But before getting into the meat of the evaluation, we discuss in Chapter Two how our analysis fits in the context of previous efforts to examine GPS and the need for backup capabilities and highlights some considerations about how to conduct an evaluation of the need for backup capabilities. It focuses on the importance of three considerations that are part of our evaluation that we believe are critical to an evaluation: (1) the importance of linking scenarios of concern to plausible threats; (2) the importance of a balanced assessment of threat effects on PNT benefits; and (3) the importance of reflecting the range of needs for timing and positioning.

In the next four chapters, we discuss the results of the four sequential parts of the analyses. Chapter Three describes both natural hazards that can disrupt GPS and potential adversarial threats to the system, providing the context for considering responses to those threats and hazards later.

In Chapter Four, we describe and contrast two approaches to characterizing the losses that would be suffered in the event of a disruption in GPS. One follows many prior studies (discussed briefly in Chapter Two) in using top-down nationwide, sector-by-sector microeconomic estimates across the economy.[4] This is contrasted with a regionally based, sector-by-sector microeconomic estimate, built from the bottom up for important sectors by analyzing selected attacks in an attempt to understand the practical constraints imposed on attackers seeking to disrupt GPS and the varied ways individuals and also the economic system can respond to disruptions in more detail.

Chapter Five describes technical alternatives that could supplement GPS or provide increased robustness and resilience to PNT. The discussion explores the performance characteristics of these options based on open-source data and compares different systems to reported accuracy and precision ranges for different industries and applications. We include more detailed proprietary cost and performance data drawn from responses to the two RFIs from the DoT in 2016 and from the DHS in 2018 in a limited-distribution annex to this report.

Chapter Six explores the cost-benefit balance of different alternatives, examining the scale of costs based on the different practical requirements of technologies and the losses they could contribute to avoiding. Considering the cost-benefit ratio for different

---

[4] See, for example, Sadlier et al. (2017); Leveson (2015a, 2015b).

complementary or other PNT technologies is complex because most systems do not support all users.

Chapter Seven summarizes the evaluation results and explains how to consider those results in different ways. This allows readers to understand what assumptions need to be made about costs, benefits, or threats that would change the cost-benefit results of efforts to strengthen the robustness and resilience of PNT over the longer term. The chapter concludes with observations based on our analysis. In particular, we identify opportunities for federal action, including public-private partnerships, by DHS or other agencies that appear to offer cost-effective solutions.

The report also contains a series of appendices: Appendix A provides a detailed explanation of the framework for estimating damages; Appendix B has more detail on the regionally based estimates; Appendix C contains more detailed descriptions of the technical options for backups; Appendix D provides additional context for comparing public-private partnerships with more ordinary acquisition programs; Appendix E summarizes the formal U.S. commitments and obligations involving GPS; Appendix F describes in more detail the potential effects of a large solar storm on the GPS satellites; and Appendix G (available as a separate document) provides a more quantitative cost/benefit estimate by relying on the proprietary cost information from the RFIs.

CHAPTER TWO

Considerations in Evaluating Backups and Complements to PNT/GPS Systems

As we discussed in Chapter One, the issue of the benefits and vulnerabilities of PNT and GPS systems is an enduring concern for both the public and private sectors because of what it enables society to do and because of how ubiquitous the systems are. Given that, it is not surprising that many other researchers have examined PNT and GPS over time.

In this chapter, we briefly discuss some of the past efforts to examine GPS and then discuss some of the considerations that are critical in considering any evaluation of the need for backups and/or complements to PNT/GPS systems—considerations that are key to how we have structured our evaluation.

Prior Efforts That Examined the Value and Vulnerabilities of PNT Systems

The value of PNT, and its potential vulnerabilities, have been the subject of a diverse array of authors throughout the public and private sectors. Various federal agencies throughout the years have either directly produced, or commissioned, a wide array of literature on PNT and specifically on GPS. These studies have included technical analyses of various alternatives or backups to GPS to cost-benefit assessments of different systems.[1] Often times, these studies are mandated to consider either alternatives to, or the vulnerabilities of GPS, as their raison d'être.[2] These studies help in understanding the consequences of a failure of GPS and assessing the position, navigation, and timing needs of different stakeholders.

---

[1] For examples of government assessments of GPS, see John A. Volpe National Transportation Systems Center (2001, 2009); Wallischeck (2016); Leveson (2009); Halsing, Theissen, and Bernkopf (2004); Cobb (2002); Roskind (2009); Eyer and Corey (2010); U.S. Army Corp of Engineers, Institute for Water Resources (2009).

[2] See, for instance, GAO (2013b); DHS, Science and Technology Directorate (2016); Wallischeck (2016); Martinez et al. (2017); Doherty (2011).

Studies conducted by the private sector are much more diverse in their form. Some studies are interested in the potential vulnerabilities inherent in PNT systems writ large.[3] Other papers are concerned with exploring the benefits attributable to GPS/PNT and the cost(s) of its unavailability.[4] For instance, Leveson (2015b) evaluates more than one hundred different studies in both the civilian and private sectors that consider the economic benefits of GPS.

Considerations in Evaluating Potential Backups or Complements

The question of potential backups or complements to GPS may seem like a simple one: estimate the likelihood of failure or degradation in function, calculate the cost if such incidents occur, and then compare the incident costs to the potential costs and benefits of complements or backups.

However, as we discussed in Chapter One, because GPS is not the sole source of PNT for many users—it is actually part of an ecosystem—the implications of GPS loss or degradation can vary significantly. Thus, the question is more appropriately framed as the value that an industry or other users get from the extant PNT ecosystem and the implications of removing or corrupting GPS's contribution to that ecosystem, *given current and potential future changes in it*. That broader perspective is critical to appropriately weighing the addition of backups or complements to the system and whether doing so is the most appropriate response to the relevant threats.

As a result, in structuring our analysis, we have made judicious choices, both with respect to threats and how to analyze the costs associated with those threats and, therefore, the potential benefits of investments in response; those choices distinguish our analysis from previous analyses in three important ways.

The Importance of Linking Scenarios of Concern to Plausible Threats

First, one challenge in assessing threats that could affect the functioning of GPS or the PNT ecosystem overall is that sources of damage or disruption fall across a wide range and that some of the more important sources are difficult to analyze with any real level of accuracy or precision. There are both "high end" threats that could have significant impacts on the functionality of GPS and "low end" threats like nonstate actors or other individuals interfering with signals locally. Estimating probabilities and consequences across this spectrum is not possible. Although large-scale space weather events could potentially affect GPS signals, such events are fortunately quite rare and unpredictable.

---

[3]   Michael (1999); Coffed (2016); Engler et al. (2012, 2016); Jiang, Zhang, and Teng (2018); McCallie (2011); Kacem et al. (2017); Yang et al. (2014); Balduzzi, Pasta, and Wilhoit (2014); Riahi Manesh and Kaabouch (2017); Purton, Abbass, and Alam (2010); Costin and Francillon (2012).

[4]   Studies in this vein include Abt Associates (2017); Leveson (2015a); North American Electric Reliability Corporation (2017); Pham (2011).

Extreme national security scenarios such as the detonation of many nuclear weapons in space could threaten the GPS constellation, but we are unable to assign a probability to such an event in a given year—in any case, such extreme military scenarios are outside this report's scope. While smaller-scale interference with GPS is more common, objective data on the actual frequency and effect of such events are limited. This limitation pushes analyses to more subjective approaches, which raise concerns about validity and reproducibility.[5] *Reflecting this reality, when we assess human-made threats to GPS, we do so without making explicit estimates of their probability to avoid the risk of creating an impression of artificial precision where none exists. We are complete in that our analysis is inclusive in the categories of potential threats, but we seek to balance the discussion appropriately to not allow individual—possibly very low probability—scenarios to dominate the analysis.*

Some past analyses have simply posited specific threat outcomes—for example, the GPS constellation is made inoperable for a specific number of days from an unspecified cause—and have used these threat outcomes as the basis for assessing or estimating the total economic benefits of GPS or PNT. Doing so essentially involves considering a scenario where all the benefits from GPS disappeared in their entirety.[6] Although using such abstract scenarios may be useful in purely technical studies, we do not believe that such approaches are best suited for policy analysis. Positing an outcome independent of the events or actors that produce them can create systematic biases in examining both the probability and consequences of these events, even in cases like this where risk analysis must be done subjectively because of limits in data.

- Assuming a set and serious outcome, such as a month-long global denial of GPS, without considering the capacity of natural events or strategic actors to produce that outcome, risks may be exaggerated in the ability of such actors to produce such a high ***consequence*** incident.
- Where the concern is adversaries—whether states or nonstate actors—causing GPS disruption or corruption intentionally, not addressing such actors' potential goals and preferences risks overstating the ***probability*** of attacks on these systems. While disrupting GPS for military purposes or seeking to cause chaos in

---

[5] For example, in DHS's 2012 *National Risk Estimate: Risks to U.S. Critical Infrastructure from Global Positioning System Disruptions*, qualitative scoring approaches were used to weigh different threat scenarios and their potential consequences. In its review, the GAO raised concerns that the study did not meet key requirements for risk analyses, including being "(1) complete, (2) reproducible, (3) defensible, and (4) documented so that results can contribute to cross-sector risk comparisons for supporting investment, planning, and resource prioritization decisions" (GAO, 2013a, p. 12).

[6] For example, the 2001 Volpe Center report discusses a vaguely defined "long term outage" with no explanation of what that entails (John A. Volpe National Transportation Systems Center, 2001; Wallischeck, 2016). The London Economics report *The Economic Impact on the UK of a Disruption to GNSS* proposes a five-day outage for the sole reason that other contingency plans for infrastructure in the United Kingdom propose five-day events. See Sadlier et al. (2017).

society by jamming the system locally is certainly an option to adversaries, it is only one of many options available to them. Even qualitative assessment of probability requires reflecting the factors that would push adversaries toward choosing such an attack **and** those that would push them away and toward more traditional military, terrorist, or criminal options. *As a result, in considering threats to GPS, we should not divorce the potential effects on PNT from the actors that cause them, instead always coupling consideration of what might happen with the why and how involved if an adversary is producing that outcome.*

Second, in considering threats to GPS functionality, those threats should not be assessed in isolation; in other words, assessors should not set aside the potential of threats to affect other components of the PNT ecosystem and potential backups or complements to GPS. This is most critical in considering adversary behavior: If a state or nonstate actor executes an attack to disrupt GPS with the goal of denying PNT to the United States or some portion thereof, assessing such an actor's propensity to target other parts of the PNT ecosystem—including systems designed to back up GPS—must be considered.[7] It is simply not prudent to assume a purposeful adversary that could target GPS but not assume the adversary is sophisticated enough to consider and, *if possible*, target other systems whose presence would otherwise make their attack on GPS irrelevant. Further, not including such threats to other PNT capabilities could create a dangerous situation where the nation would think itself protected but where the nation would in reality still be vulnerable. Such a situation could discourage individuals or firms from taking more decentralized action to increase PNT resilience. *As a result, in considering threats to GPS, we also consider the vulnerability of other parts of the PNT ecosystem to those same threats.*

For the largest-scale threats, considering "what else the incident might break beyond GPS" also suggests a logical way to bound thinking about policy choices that would otherwise be framed as narrowly focused on PNT requirements and functioning. For example, considering large-scale attacks in space—a scenario that could create a large-scale and long-term GPS outage—such an attack would likely have consequences that go far beyond whether space-based PNT systems are still functional. Indeed, the GPS constellation of 31 operational satellites, in medium earth orbit at about 20,200 kilometers (km) altitude, is a particularly difficult target; the satellites are much harder to reach than those in low earth orbit, and many satellites must be removed before the

---

[7]   A corollary to this principle is that the exposure of other components of the PNT ecosystem (including systems intended to back up GPS directly) to *other* risks must also be considered. Although intense space weather events is a threat that has been of significant concern with respect to GPS and in terms of the potential value of ground-based systems as backups that would be less vulnerable to such events, any ground-based system would be vulnerable to terrestrial weather, meaning that it would be exposed to risk that space-based systems would not. Any analysis of a value of a GPS backup or other addition to the PNT ecosystem must not ignore such risks, which at the minimum would represent a contributor to the operations and maintenance costs of such an alternative system.

PNT functionality would be significantly degraded (GPS.gov, 2019). GPS is unlikely to be so attacked unless many other satellite systems have already been neutralized, as well as broader, nonspace attacks. Put simply, in circumstances involving significant space attacks or even a nuclear exchange (and the other actions that would likely be taken by any state adversary in such a scenario), the loss of PNT functioning would be unlikely to be highest on the list of national concerns; therefore, investments to hedge against PNT loss would be unlikely to be prioritized over ones to address other consequences of such a conflict.

Reasonably foreseeable responses within the logic of the scenario are potentially important as well. For example, in a military engagement where an adversary acted to deny GPS so definitively, the DoD would have incentives to reconstitute capability as rapidly as possible, and civilian and other sectors would be able to benefit from that drive to restore PNT. Similar arguments apply to cyber threats to GPS systems. For example, in a scenario where a state adversary targets the system through cyber means, the adversary would also likely target alternative sources of PNT for logical consistency; once again, there would be similar military capability drivers to restore or reconstitute capability rapidly. *As a result, in considering the largest-scale threats to PNT, maintaining this larger context for the analysis is important to avoid narrowly focusing on PNT-specific effects in scenarios where other considerations may and should realistically dominate.*

The Importance of a Balanced Assessment of Threat Effects on PNT Benefits
The starting point in considering the benefits of acting to strengthen the PNT ecosystem is how serious the consequences of threats to PNT are in reality. Since the effect of a threat scenario is to deny users the efficiency and effectiveness benefits they gain from GPS, the scale of the threat is defined by how large those benefits are, which differs considerably from industry to industry. Making those comparisons requires a common economic framework, including the ability to monetize benefits to life safety that can be difficult to objectively value. Much of the prior literature on the benefits of GPS, both here and in Europe, has relied on nationwide, sector-by-sector microeconomic analyses (e.g., Sadlier et al., 2017; Leveson, 2015a, 2015b). *Such analysis is well matched to exploring large-scale effects of disruptions to GPS; thus, to maintain consistency with prior literature, we do such an analysis as a central part of our assessment.*

However, this level of analysis is less well matched to looking at more time-limited events like jamming and exploring how such incidents might be resolved by users by using existing complements or other adaptive responses. This is an important limitation of nationwide analyses because short-term jamming and other more localized events are a key part of the threat space for GPS. Being able to place such smaller-scale events in time is also important because the effects of GPS disruptions may vary day to day within an industry. For example, the effect of disruption in precision agriculture will depend on the time of year and specific day, with the greatest impact of disruptions occurring at key points like planting and harvest. Fortunately, the specific days of import vary with crop and, within a given crop, by latitude, so only a small fraction of the value from precision agriculture would be lost during a short outage. Even if a short-term disruption is national in scale, delaying tasks until functioning returns can blunt the effects of the disruption. Whether complements or local backups to GPS are already part of the PNT ecosystem is also important for smaller localized events. As discussed above, examinations of threat must assess if those other components of the ecosystem are likely to be disabled or degraded in a scenario of concern. However, the converse is also true: If they are *not* affected and can thus mitigate impacts from the loss of GPS, assessment must reflect their value and the contribution they make to resilience to GPS disruption in any benefit calculation. The availability and relevance of different PNT solutions to different industries and users varies; thus, appropriately reflecting that reality requires focused analysis. *We use nationwide analysis as well as regional analyses of specific geographic areas and affected industries, making it possible to look at incidents in more depth and to better analyze existing backups and workarounds.*[8]

Just as logical consistency in threat analysis requires that we make reasonable assumptions about the capabilities of potential adversaries, we must also take a realistic view of the incentives and capabilities of those who deal with GPS-dependent systems. The first way this shapes our analysis is in thinking about the potential for a GPS outage or disruption to have large and unbounded cascading effects. For a technology that touches so many infrastructures and where an outage would be noted and experienced by so many people, one might assume that any incident would have effects that would ricochet across the economy and greatly magnify its effect. While this is true at a basic level—the effects and costs of a GPS outage *are* driven by how PNT is used by others—past experience suggests that disruption or spoofing of the signal would not inexorably result in catastrophic failures or widespread damage in other systems.[9] This result is not that surprising: Looking at infrastructure as a prudent engineer would, there are good reasons to protect against a variety of challenges, thus allowing for some resiliency in the system.

For example, it has been posited that jamming or spoofing of GPS could cause physical damage to power grid systems directly under particular circumstances, but we do not include those effects in considering "threats to GPS" and thus do not include the potential for such catastrophic or cascading outcomes on the "potential benefits ledger" for future investment in PNT systems. This is because we see addressing these issues as primarily a system engineering concern at the level of the *use* of PNT signals, where it is reasonable to assume that powerful incentives exist for infrastructure owner-

---

[8]   Workarounds defined as methods to bypass or overcome a potential GPS outage.

[9]   See the section "Some Historical Evidence Against Predictions of Extreme Disruption" in Chapter Four for more detail.

operators to implement available solutions for systems where there is the potential for serious or widespread damage.[10] Given the existing PNT ecosystem, there are alterative and complementary sources of PNT that are already available to address the risk. If the owner-operators of such vulnerable systems have not taken advantage of the existing options to respond to threats of spoofing or signal loss, it is unclear how adding one or a small number of backups or complements to this system would significantly change their incentives.[11]

Extending this argument, in considering the economic and other effects of threats to GPS, we do not see *spoofing* GPS signals as more significant than their *jamming* for nontransportation cases in this consideration of alternative PNT systems. The rationale for that conclusion is that, given the ability to readily detect spoofing in sensitive infrastructure applications by using integrity monitoring, capabilities already exist to limit its potential impact. For transportation systems, adversaries have been shown to spoof GPS signals in both military and civilian applications, for example making it possible to alter the course of individual vehicles that are relying on GPS positioning to guide their movement. However, integrity monitoring and related cross-checking technologies to guard against corrupted information *are* available to mitigate much of the difficulty in the most sensitive applications where safety of life is at stake. While such integrity monitoring may not, in fact, always be implemented, that is a problem in system design or regulation, and not due to the absence of an alternative.

As discussed above for infrastructure, solely relying on GPS signals for positioning in cases where an error would cause large damages or loss of life would be both unlikely and unwise, given existing opportunities for integrity checking. It would either represent poor system engineering or the acceptance of a level of risk that would not be typical of the actions of most system designers. Also, as above, simply providing another alternative positioning or timing signal would not address such an engineering/design.[12] Indeed, in our research on using GPS in particular industry sectors, many who *use* GPS do not solely *rely* on it, therefore enabling significant adaptation even in scenarios where GPS signals would have to be treated as suspect in their accuracy (e.g., GPS integrity monitoring to alert system designers when disruption or corruption

---

[10] In this we therefore diverge somewhat from the elicitation findings of the National Risk Assessment, which does include assumptions that this sort of damage will occur (e.g., loss of timing in the electrical grid leading directly to overloading and disruption).

[11] Note that other types of interventions (e.g., regulatory requirements) might change their incentives.

[12] Indeed, even if provision of a new alternative PNT signal was chosen as part of an effort to address spoofing as a threat, its implementation to solve this problem would be a massive system engineering challenge in and of itself, requiring changes in technology and practices by PNT users to actually utilize the signal for this purpose. It does not seem credible to assume that simply providing another alternative signal would materially change the incentives for users to make such changes where they do not exist now to adopt existing options (e.g., utilize multiple GNSS signals simultaneously, combine use of inertial measurement units [IMUs] with PNT signals, or alternative timing systems).

occurs). Put simply, for both infrastructure systems and vehicles, if GPS dependence has been built into systems to produce such extreme vulnerability, public investment in additional or backup PNT capacity seems less appropriate than focusing on addressing the design incentives that allowed the vulnerability to arise in the first place.

The Importance of Reflecting the Range of Needs for Timing and Positioning

If policymakers want to consider national policy toward GPS and PNT more generally, they must understand the different ways that these technologies are used across the economy. Different requirements for precision and accuracy drive what types of backups and complements are acceptable. Across these different sectors and applications, there are different requirements for real-time PNT to support dynamic activities given the potential value of stored legacy data collected and recorded at high precision and accuracy. While the timing component of PNT is an inherently real-time application, positioning (and some navigation) tasks can be supported by legacy data. For example, while real-time positioning is necessary for dynamic turn-by-turn navigation, mapping applications where GPS or other PNT systems have been used to develop precise static maps do not require real-time positioning.

As a result, some applications in a sector may be more sensitive to disruptions in GPS or other PNT systems than others. For example, within agriculture, precise piloting of farm equipment might be affected by PNT disruption, but mapping of farmland would not be so affected. Of course, in the former case, piloting of farm equipment could still be done manually, albeit with less precision and thus some inefficiency. Previous work by others has explored the PNT needs of users in critical infrastructure sectors. Cavitt et al. (2018) surveyed users for their precision timing requirements while Tralli et al. (2018a, 2018b) identified user needs for positioning and navigation. We will return to the specific values developed in others' work on variation in accuracy and precision across sectors in subsequent chapters of the analysis.

Additionally, we investigated in some depth the existing backups and workarounds for any disruption in GPS. As our economic analysis in Chapter Four demonstrates, understanding such backups and workarounds makes a difference.

CHAPTER THREE

Identify Threats and Vulnerabilities of GPS and Other PNT Systems

In framing the potential benefits of investing in adding additional backups or increasing the resilience of the PNT system, a clear point of entry is understanding the nature of the hazards those PNT systems face. By reducing losses from hazards to current PNT functionality and stability, individual backup and resilience options can be beneficial for different types of threats. However, as framed in Chapter Two, the exposure of those backup and resilience options (both existing and potential future additions) must also be considered. As a result, policymakers must understand the nature of the risks to the functionality of GPS **and** to any other complementary elements providing PNT; doing so is the starting point for considering the economic and policy rationales for investing to strengthen the resilience of the PNT ecosystem.

In this chapter, we address the first of the four analysis parts from Table 1.1: *identify threats and vulnerabilities of GPS and other PNT systems.* That part of our analysis focuses on answering the following questions: What are the potential threats to GPS, from nation-states, subnational groups, and natural events? How large an area is affected from each threat, how severe is the disruption, and for how long would the disruptions last? And what can we say about the probability of such events? *Our bottom line answer to this set of questions is that policymakers need to think systematically about all the plausible threats, not just the larger threats that dominate policy discussion. When we do so, we find that the plausible threats to GPS are limited in duration and/or area: major solar storms only last for a few days, and it is not plausible that enemy jammers could affect large areas of the continental United States (CONUS) for a long time.*

The remainder of this chapter highlights how policymakers should think about the risks from this full range of threats systematically using a conceptual matrix that is oriented around intention and scale of the attack; it then shows the results of such a systematic examination. It also examines one specific threat that is outside the matrix—cyberattacks on GPS.

A Systematic Way to Consider the Risks of a Full Range of Threats

In the existing policy and technical discussion surrounding PNT and the policy debate surrounding GPS in particular, large-scale natural events (notably, the effect of space weather on satellite systems) have been a central focus. However, a range of different events and actors can affect the functioning of satellite and terrestrial PNT capabilities. For example, there have been well-documented incidents of individuals using GPS spoofing or jamming technologies for their own purposes and of events where strong electromagnetic transmissions accidentally disrupted PNT signals.[1] This means that in thinking about the full range of threats, analysis must reflect all of the following items:

1. Large-scale threats that result in the loss of PNT capabilities by the *destruction* of systems
2. Smaller-level threats that lead to the *denial* of those PNT capabilities through jamming or other types of disruption
3. Smaller-level threats that lead to *spoofing* signals to fool PNT systems into accepting incorrect data.

To consider these threats systematically, we break them down into categories, distinguished by scale and intentionality (i.e., whether the threat comes from strategic actors who are seeking to consequentially disrupt PNT or whether the disruption is from natural or unintentional sources). For discussion, this allows risks to be binned into four categories, as shown in the matrix in Figure 3.1, which includes illustrative examples of such threats in each cell. We use these four categories to discuss the types of effects that each threat could have for PNT, including GPS and potential complements or backups to GPS; the discussion will serve as input to inform analysis of the economic effects these threats could produce—something we discuss in Chapter Four.

In each of the four categories, there are also important differences in the likely timing and duration of threats and, therefore, in the ability of users who rely on GPS to respond to them; such responses include both technological ones and behavioral and other responses as well. Natural and negligent threats will arise randomly, though not necessarily wholly unpredictably; therefore, their effects will depend on how their occurrence coincides with high value PNT use in the areas where they occur. In contrast, strategic actors will seek to target their actions to produce enough of an effect to achieve their goals, though whether they can do so optimally will be driven by their ability to gather the information needed to do so.

In considering the different types of threats to PNT, these categories also assist in framing *plausible* and *realistic* threat scenarios. As discussed previously in the framing of our analysis, exploring individual outcomes in isolation can be done for analytical

---

[1]   For example, incidents in Half Moon Bay, California, in 2001 and in San Diego, California, in 2007.

Figure 3.1
PNT Threat Matrix

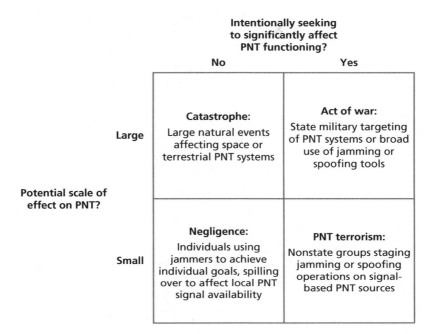

**Intentionally seeking
to significantly affect
PNT functioning?**

	No	Yes
Large	**Catastrophe:** Large natural events affecting space or terrestrial PNT systems	**Act of war:** State military targeting of PNT systems or broad use of jamming or spoofing tools
Small	**Negligence:** Individuals using jammers to achieve individual goals, spilling over to affect local PNT signal availability	**PNT terrorism:** Nonstate groups staging jamming or spoofing operations on signal-based PNT sources

**Potential scale of
effect on PNT?**

purposes (e.g., positing the total loss of the GPS constellation as a result of an event of unspecified characteristics); however, while such an approach has the analytical advantages of simplicity, it could produce sufficiently misleading results that do not justify such advantages.

Policy decisions surrounding investments in PNT must be informed both by the abstract value of the capability they seek to protect and by the circumstances that could lead to its loss; this is because those same circumstances could affect the relative benefits of the options that are available as well. Further, for scenarios involving human adversaries—whether state actors, terrorists or criminals—details matter. Situational specifics can enable or constrain the ability of such actors to act, and the intent, capabilities, and goals of specific actors shape the threat they pose both to PNT and to potential alternative systems designed to strengthen the resilience of the PNT ecosystem.

Explicating the Threat Matrix

As noted, the matrix in Figure 3.1 contains four categories: catastrophe, act of war, negligence, and PNT terrorism. Here, we discuss each category.

Category 1: Catastrophe

In the large-event category, major natural events are the most obvious concern for the functioning of the PNT system and for GPS in particular. Space weather certainly holds the potential for solar events to disturb the ionosphere for days, significantly degrading the performance of GPS and all GNSS systems. We discuss extreme space weather events and then other large-scale natural events.

Extreme Space Weather Events

The source of space weather events involves activity occurring on the surface of the sun. A solar flare—defined as an explosion at the sun's surface (the photosphere)—generates a burst of electromagnetic radiation, including intense radio waves, X-rays, and gamma rays; such radiation travels outwards at the speed of light, reaching earth in approximately eight minutes. The radiation bursts can heat the earth's upper atmosphere, increase ionization, and can generally lead to an enhancement of electrical currents and electron density in the ionosphere. The solar flare event can also generate solar energetic particles (SEPs), particles traveling at a range of relativistic speeds, which can reach the earth system in tens of minutes to a few hours after the flare. These also contribute to the heating and ionization of the ionosphere. This changes how satellite radio transmissions refract through the atmosphere, disrupting high-frequency communications signals. For GPS, this "scintillation" in the upper atmosphere is a common cause of "loss of lock," where ground-based GPS user equipment loses reception, making GPS ineffective for periods of tens of minutes up to a few days, depending on the duration and recurrence frequency of the solar events.

Another space weather effect comes from large ejections of a mass of plasma by the sun. In a coronal mass ejection (CME), which is sometimes coincident with a solar flare, the bulk of solar plasma material expelled from the sun takes significantly longer to travel through space and is not always aligned with the earth's azimuthal position in its orbit around the sun. However, when it is aligned, it can reach earth in anywhere from 14 hours to four days, depending on the variable speed of material from the event. The effect at the earth depends on the CME speed and the orientation of the interplanetary magnetic field in the CME, which can help or hinder the injection of material into the magnetosphere through magnetic reconnection. Typical results of major CME-driven geomagnetic storms are particle precipitation at the magnetic cusps (visible as aurora); injection of high energy particles into the ring current and radiation belts; and thus for satellites, increased radiation, a larger total ionization dosage, and higher probability of single event effects depending on orbital altitude (Royal Academy of Engineering, 2013; see Appendix F for a discussion of the likely very limited effect of such an event on the GPS satellites themselves). The largest such events also heave the entire ionosphere, thereby generating large induced currents in long conductors, such as power lines (Odenwald, 2009); for GPS and other GNSS, this means further perturbing their signals.

The Carrington event of September 1–2, 1859, is often referenced to anchor the extreme high end for effects from solar events. The event was documented when English astronomer Richard Carrington directly observed a major solar flare through a solar telescope. Seventeen hours later, a geomagnetic storm was observed through the deviation of magnetic needles at the magnetic observatory in Bombay, India. These magnetic disturbance deviations have been translated to a storm disturbance index (Dst) of −1760 nanotesla (nT).[2] Intense auroral activity was observed in both hemispheres, with reports of the aurora being visible as low as 23 degrees magnetic latitude; in addition, several telegraph lines were overloaded because of induced geomagnetic currents (Tsurutani et al., 2003). These parameters place the Carrington event as the most extreme geomagnetic storm to occur in the past 159 years. Analysis by others has concluded that the Carrington event was a solar flare coupled with a CME of solar plasma that traveled at an extremely high speed through interplanetary space. Once the CME reached near-earth space, it circulated through the magnetosphere, resulting in the observed geomagnetic storm, which lasted less than two days.

Of course, there were no artificial satellites in space or radio signals during the Carrington event. However, there have been many major CMEs and other space weather events since then, both before and during the space age, and these events have helped empirically demonstrate some components of the impact on space-based systems.

Estimating the likelihood of such extreme events occurring again is obviously difficult, as the historical record is short. The Royal Academy investigated this, and concluded that "various studies indicate that a recurrence period of 1-in-100 to 200 years is reasonable" and "we anticipate that a solar superstorm might render GNSS partially or completely inoperable for between one and three days" (Royal Academy of Engineering, 2013). Riley (2014) has performed statistical calculations based on over 50 years of records of CME-driven geomagnetic storms and has estimated that Carrington-like superstorms are likely 100-year events, with an approximate 12-percent probability of occurrence in a given decade. That said, less extreme solar flares and CMEs that may be significantly geoeffective are more frequent. Hence, the probability of these kinds of events occurring is significant, and owners/operators of space-based systems should be prepared to deal with them. For the purposes of the later economic analysis, we use a conservative estimate, of a 100-year repetition rate and a three-day GPS (and all GNSS) outage due to an extreme solar storm.

Fortunately, there are established methods of coping with the effects of a solar storm on spacecraft, from shielding or otherwise hardening electronics to "safing" systems during the actual event by suspending operations when the spacecraft is passing through a region or time of expected inundation by higher energy particles that also

---

[2]   A negative Dst value means that Earth's magnetic field is weakened. For reference, anything below −100 nT is considered a significant geomagnetic storm.

interrupts whatever service they provide. However, for satellites like GPS that reside in an ambient radiation belt, the effects of even a Carrington-scale event on the satellites appear to be minimal (see Appendix F for more details).

Overall then, we find that the largest effect of a solar storm on the GPS system would be the distortion of its signals when transiting the ionosphere because of the scintillation induced by the storm. As noted, for our analysis in a Carrington-scale event, we adopt a conservative, three-day estimate for the duration of this disruption, assuming for simplicity that no GNSS service is available at all during this period.[3] However, an event large enough to have significant effects on GPS signals could also have effects on terrestrial power and communications systems (through the heaving of the ionosphere as in the Carrington event's effects on telegraph systems). As a result, other components of the PNT ecosystem that rely on those infrastructure systems could also be damaged or disabled by such an event.

Other Large-Scale Natural Events

Although space weather events have been the primary focus of concern about risk to PNT, large-scale natural events could also affect the ground-based components of GPS. Such large-area events would also pose risks to ground-based alternatives to GPS seeking to increase the robustness of resilience of the PNT system overall. Hurricanes, tornadoes, or earthquakes could damage sites or transmission towers that GPS and terrestrial alternative systems rely on.[4] Local positioning solutions, such as those at a port, could be affected by even smaller natural or accidental events, such as localized flooding or power loss.

Large-scale terrestrial natural events are much more frequent than space weather events. As a proxy for the largest and most consequential such events, the National Oceanic and Atmospheric Administration (NOAA) tracks natural disasters costing or producing total damages of more than a billion dollars. Such events have occurred multiple times per year, affecting subsets of the country, with many having effects that could disrupt power and infrastructure-dependent, terrestrially based PNT systems (NOAA, 2017).

Category 2: Acts of War

Acts of war is a second category of threat shown in Figure 3.1. Dealing with acts of war is primarily the responsibility of the DoD; national security threats writ larger are formally outside the scope of this report. To the extent DHS must be concerned about such scenarios, its responsibilities usually focus on consequence management. Since

---

[3] It is also the case that a solar event with sufficiently large effects to materially affect GPS could also have terrestrial impacts on power and other technological systems.

[4] For example, Celano, Peterson, and Schue (2004) describe damage to Loran-C towers as a result of hurricane impact. Other types of natural events (e.g., stress from icing or snow, other extreme wind events) also have caused damage or failure of broadcast tower infrastructures.

war is a plausible path to a large-scale effect on GPS/PNT functionality, we discuss a range of possible scenarios and their consequences to frame this portion of the GPS threat space.

Acts of war that can have a large-scale effect on the functioning of PNT systems are a subset among a much larger possible variety of military scenarios. While PNT is a legitimate military target, attacking it in a way that denies its utility globally, or to North America, would be essentially infeasible for any but the most advanced state adversaries; we believe that Russia or China alone are in principle capable of either the comprehensive cyberattacks required (discussed separately below) or physical attacks on numerous, radiation-hardened satellites in medium earth orbit (MEO).[5] In addition, such targeting in space could cause collateral damage to those countries' own systems; even if that does not happen, such an attack would risk in-kind attacks by the United States on their space assets or strikes against relevant sites in their homelands. Further, from a military planning perspective, targeting GPS as a constellation to deny the U.S. PNT capabilities would, at best, be an indirect way of affecting the U.S. military in a local military theater. Choosing to carry out such strikes in space would require using either a large number of nonnuclear antisatellite systems capable of reaching MEO or a significant number of nuclear weapons. The latter would bring in a new list of concerns about escalation. Additionally, any nuclear attack in space powerful enough to damage radiation-hardened satellites might also result in other collateral effects (e.g., electromagnetic pulses strong enough to affect systems on the earth's surface), meaning even a space-focused attack might damage electricity-dependent terrestrial backups or GPS complements. Somewhat similarly, attacks (presumably conventional) against the ground infrastructure of GPS would offer only a slow degradation of GPS capabilities, and because these would be attacks within the United States, they would also raise concerns of escalation.

In comparison, attacks against PNT use in regions of direct military interest would appear to yield a predictably greater return per unit of investment since the effects would be direct and immediate. These theater attacks, including terrestrial spoofing and jamming of GPS systems, would be integrated into more conventional military plans and would likely serve as a support element of a broader range of activities. As a result, we conclude that even though attacks on GPS in the domestic United States cannot be ruled out, doing so would unlikely be a high priority target in large-scale great powers conflict itself. If a conflict is posited in which targeting the GPS system to deny PNT domestically **is** a primary adversary goal, logical consistency requires acknowledging that an adversary capable of pursuing that goal would also include targetable PNT alternatives in their attack plans.

Beyond scenarios involving the physical destruction of PNT assets and disruption or spoofing during an active conflict, jamming and spoofing systems could also

---

[5]   See Zenko (2014) for a review of space attacks.

be applied by nation-state adversaries within the geographic boundaries of the United States. In these sorts of smaller and less visible attacks, the nation-state would be trying to conduct largely deniable operations to minimize the risk of military response while still causing some harm to the United States. In this case, such attacks will share many of the characteristics of the nonstate actors described below as PNT terrorism, though the mix of resources and motives likely differ in detail. Compared with the investments required for warfighting, the investments and resources needed for these sorts of attacks are smaller and the risks following detection are smaller; but as we will show, the gains are also small.[6]

It is important not to assume that theater military experience maps to domestic U.S. society; the situations are very different for the attacker, and the range of operations designed to be interfered with are quite different than those seen in the civil and commercial world in the United States. While U.S. military forces are hardened against many attacks compared with civilian users, the scale and types of attack assets devoted to them are enormously greater than what could feasibly be applied to the U.S. homeland. Also, in an active military conflict, even brief denials and spoofing of PNT might make a difference if well timed with other operations.

Brief denials or spoofing in a civil economy are quite different. In a broad economy, time and the wide varieties of processes in play make it difficult for the attacker to get predictably good results. Further, the proliferation of alternative positioning signals—both GNSS-based systems and other alternatives—means that jamming or spoofing attacks could be more difficult to carry out in a civilian context than in a military conflict. Some modern GPS receivers—as those in common cell phones—now also use the Russian GLONASS constellation and will soon use the European Galileo GNSS; this is because the chip-scale integration for receiving the signals and the needed software are widely and cheaply available. While using a PNT signal provided by a potential adversary system might be viewed as risky in a military engagement, the risk calculus would be different for civilian applications. Additionally, the location information from GNSS is combined within modern cell phones with other sources of location information, such as cellular signals. As a result, for jamming or spoofing approaches to be effective, other alternative systems that commercial devices are already utilizing would have to be disrupted as well, thus increasing the difficulty of achieving predictably large damages or disruptive effects.

---

6   In the DHS's National Risk Assessment, which included expert elicitation of predicted duration and extent of disruption caused by a variety of GPS jamming and interference scenarios, most jamming scenarios were estimated to produce intermediate levels of disruption on the order of a week or less (in large part driven by the assumed time to detect and locate the sources). The exception was the use of multiple low power stationary and mobile jammers, where they estimated that the effects could be maintained for greater than a month due to the ability of attackers to remain at large and continue operations. Use of spoofing was thought to potentially produce more consequential effects (potentially by directly causing damage to infrastructure systems) for extended periods.

Category 3: Negligence

Negligence is another category in the matrix, one that occurs when individuals use GPS jamming and spoofing technologies for their own uses but also inadvertently disrupt users nearby. Although the use of the devices themselves is intentional, their broader effects on others beyond the user are not; thus, we have labeled them *negligent threats* to PNT functioning. GPS jamming and spoofing technologies are available on the open market, and there are individual incentives for using them to evade GPS tracking (e.g., by commercial truck drivers; see Brunker, 2016) or spoof positions (e.g., to cheat in augmented reality gaming).[7] Using such devices creates the potential for incidental jamming or disruption of PNT for users in nearby locations. Such jamming sources are low power and therefore generally constrained in the area they effect. This category also includes wholly unintentional jamming of GPS—and, thus, definitively negligent threats—from emitters that are incidentally transmitting in or near the relevant spectrum bands.[8] Such incidental jamming can be higher power and affect wider areas.

Because these attacks are not designed to be broadly disruptive, they have very different characteristics than threats created by strategic actors. Active efforts to evade detection will be unlikely to be used (or used effectively even if attempted), making it easier to detect and apprehend rogue transmitters. In the presence of some level of capability to locate such transmitters (e.g., locally placed or deployable detection capabilities), higher-power negligent threats—which could affect broader areas—would be expected to be detected and identified relatively rapidly. The ability to detect and shut down such transmitters would limit the duration of their effects, regardless of whether those effects are very local (for negligent use of privacy jamming or short-range spoofing) or broad (for higher-power incidental jamming).

Individual deployment of these tools would likely produce specific geographic dispersion and movement behavior. Use of GPS privacy jammers by individuals in the transportation sector (e.g., truck or delivery drivers) would produce a localized PNT disruption moving along road networks (e.g., a truck going through a major highway corridor, a delivery vehicle following a preplanned routing through an urban area). As such, the disruption would behave similarly to a localized weather event where the disruption and its costs would change based on the infrastructures or economic activities that were in its path. Travel of negligent disruptors in corridors passing by

---

[7] Most GPS spoofing attacks for gaming appear to have involved software-based methods that affect only the single intended device. Transmission-based spoofing of signals has been demonstrated, however, and if games implement changes that make software-based spoofing difficult, there may be greater incentive for users to adopt transmission-based methods. See discussion in U.K. Government Office for Science (2018).

[8] An example of such a larger scale negligent threat was the 2007 incident resulting from a U.S. Navy communications jamming exercise in San Diego Harbor. In that case, the absence of detection and response capability meant that the jamming was not rapidly detected (DHS, 2011).

multiple critical infrastructures could have diverse (if transient) effects;[9] travel through rural areas might only have small and transient effects on agriculture and other nearby drivers using GPS. Deployment of detection devices both in the United States and Europe have shown multiple—though still relatively small—numbers of daily incidents of GPS disruption assumed to be associated with negligent behavior by individual jammer users (U.K. Government Office for Science, 2018; DHS, 2011, 2012).

This threat to PNT becomes more consequential as numbers increase; a transient disruption from one driver per week might have a negligible consequence, while GPS disruptions from 100 drivers passing a major airport every day would be much more serious. As numbers increase, however, incentives for broader deployment of detection technologies for transmitters in GPS bands (Chronos Technology, 2018) also increase, along with the potential revenue from fines associated with enforcement efforts. In the United States, this requires changes in laws so that local law enforcement agencies can pursue such interference; at the moment, only the Federal Communications Commission (FCC) can pursue fines (FCC, 2018). Because negligent actors are using these technologies for varied types of personal gain, they would likely be significantly influenced by fines being imposed and publicized; this would make such enforcement an effective way to shape their decisionmaking and reduce risk. While such efforts are likely to be limited in scope, they would address any particularly troublesome cases of negligent jamming. Unlike threats originating from strategic actors, negligent denial of PNT through the use of privacy or other jammers is focused on GPS; thus, complements that operate through other technologies or frequencies will likely not be affected.[10]

Category 4: PNT Terrorism
The final category in Figure 3.1 is something we have called *PNT Terrorism*. As was the case with acts of war, such PNT terrorism would be taken with the intent to disrupt GPS or PNT functioning, but it would be staged at a smaller scale by either nation-states (seeking more modest effects) or nonstate actors that lack the capability for broader action. Although the term *PNT Terrorism* suggests that such attacks would have political aims, criminal groups could also stage such operations to extort money or achieve other goals that are generally more modest than what is seen in the larger-scale military conflicts. Available data on activities of terrorist groups to date shows no evidence that past nonstate groups have targeted PNT-type systems. However, given

---

[9] An example of these effects is the jamming detected at the Newark International Airport described in DHS (2011).

[10] The more such alternative or complementary technologies are integrated into the systems that individuals are seeking to jam to maintain their privacy, the greater the likelihood that the technologies will be modified in an attempt to do so. This would be more technically challenging, and would mean that the jammers would be emitting over a wider set of frequencies, potentially facilitating detection and intervention.

the commercial availability of GPS jamming and spoofing technologies, such an attack could, in principle, be staged by a nonstate group.

The goal of such an effort might be destruction (e.g., seeking to cause infrastructure incidents by disrupting PNT at key locations like airports or ports), but it might also be disruptive (e.g., trying to sow chaos in an area through PNT jamming). Although in principle such a group could deploy a large number of devices to interfere with PNT, practical constraints (which we demonstrate in greater detail in subsequent chapters) mean that the effects of such an attack would most likely be smaller than at the national scale.[11] With the ability to cover parts of metropolitan areas (or using upper-end estimates of resources available to such a group parts of a state), the extent of effects a group could produce would depend on how long a campaign of GPS disruption or spoofing could be sustained. However, as above for the military case, availability of alternative GNSS signals would mean that greatest effectiveness would require not just jamming GPS signals but somehow also interfering with alternative signals and complement systems as well.

GPS Jamming

Table 3.1 summarizes the types and characteristics of three classes of GPS jammers.

The range numbers in Table 3.1 are ranges at which the jammer should break the Civilian Acquisition or Access Code (C/A code) lock, that is, stop a receiver from tracking the civilian GPS signal that it was previously tracking. These range numbers are conservative in that they assume peak GPS satellite signal strength; the jammer might have an effect at somewhat greater ranges by breaking lock on satellite signals at less-than-peak strength.[12]

In thinking about the damages imposed by intentional jamming of GPS, this analysis makes it possible to take into account the practical constraints that affect how well such operations can achieve their intended goals. Considering the problem of operational design from the perspective of an attacker is important, as if the practical constraints on the operation are significant enough, the attractiveness of the operation could be significantly affected, given that disrupting GPS is a means to an end and not an end in itself. The central practical constraint on jamming operations is one of positioning. Where jammers are placed can have a significant effect on their effective ranges in a given scenario. To have the desired effect, the jammer must be placed so that it is within effective range of the target and ideally has good line of sight to

---

[11] A series of coordinated, nearly simultaneous attacks does not change this, as discussed in more detail when we estimate the potential damages from such attacks in Chapter Four.

[12] Not shown in Table 3.1 are the ranges at which the jammer might prevent a GPS receiver from acquiring any satellite signals when first turned on. In theory, that range could be very large, nine times the break-lock range, if not limited (as it probably would be) by line of sight. Threat source material and empirical tests of jammers, like the sources for Table 3.1, tend not to state such large theoretical ranges and instead just state ranges perhaps 50 percent greater than the break-lock range.

Table 3.1
Effective Range Depending on Jammer Power Level and GPS Shielding

System	Range to Break C/A Code Lock with Line of Sight to an Exposed GPS Receiver (kilometers)	Range to Break C/A Code Lock of a GPS Receiver with 20 dB Protection[a] (kilometers)
Vehicle-based jammer[b]	750	75
Large human portable jammer[c]	95	9
Small omni jammer[d]	4	0.4

SOURCES: Jane's by IHS Markit, "SCL-300 GPS Jammer," March 13, 2016; Jane's by IHS Markit, "Optima-2.2 GPS and GLONASS System Navigation Jammer," August 15, 2017a; Jane's by IHS Markit, "Optima-B GPS Terminal Spoofing Jammer," August 16, 2017b; Jane's by IHS Markit, "GPSJ-25 GPS Jammer," September 15, 2017c; Jane's by IHS Markit, "GPSJ-40 GPS Jammer," September 15, 2017d; Jane's by IHS Markit, "GPSJ-50 GPS Jammer," September 15, 2017e; R. H. Mitch et al., "Signal Characteristics of Civil GPS Jammers," in Proceedings of the 24th International Technical Meeting of the Satellite Division of The Institute of Navigation (ION GNSS 2011), Portland, Oreg., September 20–23, 2011.

[a] 20 decibels (dB) of protection could be provided, for example, if a ground receiver were moderately shielded by buildings.

[b] 1 kilowatt (kW) of transmit power with 30-degree × 30-degree beam (15 dBi antenna) for a total of 45 decibel-watt (dBW) Effective Isotropic Radiated Power (EIRP).

[c] 50 watt (W) of transmit power with 35-degree × 35-degree beam (10 dBi antenna) for a total of 27 dBW EIRP.

[d] 1 W of transmit power, 0 dBW EIRP.

the target so that the signal is as strong as possible. If possible, the jammer should be elevated to maximize coverage if it is attempting to cover multiple targets. Local topography and the built environment—for example, buildings in between the jammer and the intended target—are real constraints on the breadth of potential effects.

Other factors affecting the dispersion of targets in space in time can also constrain effects. Physical geography matters: It is easier for an attacker to cover a significant portion of an industry in a city if all the relevant sites are close together (e.g., a dense financial district) rather than if they are dispersed across the city (e.g., construction sites in many separated locations). Time matters as well: For some industries, access to PNT may only be needed at key points of a process or activity while others may use it all day, every day. As a result, in thinking about targeting GPS use in complex economic systems, one is immediately drawn to thinking about the particular periods of greater or lesser vulnerability to disruption and the importance in focusing on the more significant periods of vulnerability of any economic processes to the loss of GPS.

The process looks approximately like this for a given target area chosen by an attacker.

- A jammer can cover a finite area at particular points in time, and targets must be in the area defined by the interaction of the signals being put out by the jammer and receivers in a real-world terrain that might block or weaken those signals.

- The target must be in a phase where GPS has significant impact on the key related processes—that is, it must be using GPS to do something of value at the time, or loss of GPS will not result in a cost.
- If the target is vulnerable to disruption, then costs will add up for however long the jamming persists unless backups or other complements allow it to continue operations absent GPS (e.g., a target that has access to full backup PNT will only have losses if that backup fails somehow). Targets with only partial or unpracticed backups would likely experience some losses over the period of jamming.
- How long jamming persists will be driven by the rate that jammers are removed, run down, moved out of the area (e.g., for a jammer on a moving vehicle), or neutralized by effective countermeasures.

For any targeted areas, the effective number of entities that could be covered by a specific scale jamming operation (and therefore the fraction of economic or other activity that could be affected) is therefore a function of the following:

- the geographic target distribution in the area
- jammer coverage (or the geographic scale of the natural disruptive event)
- the number of jammers
- interactions with security forces and local population removing jammers or apprehending individuals placing them (where applicable).

All other factors equal, the larger the number of jammers with a given range, the greater the number of targets covered given the relatively low density of most economic activities.

Given the relatively low cost of the technologies involved commercially, we would not expect the financial resources of a group to be a strong constraint on capability when using lower power systems, though a vehicle-based, higher-powered directional jammer with tracking capabilities seen in the military arena would be more difficult to obtain and would be fairly expensive. Estimates of resources available to *major* threat groups go into the hundreds of millions of dollars, and resources invested in single conventional terrorist attacks have ranged from the hundreds to tens of thousands of dollars (Institute for Economics and Peace, 2017). Individuals unconnected with larger groups would be expected to have fewer resources, but could still likely afford multiple low-cost jammers.

The longevity of a disruption campaign in the face of investigative and response activity would be determined by a group's skills and how its deployment strategy affected the probability of getting caught at different points in its activities. As strategic actors, groups intentionally engaged in disrupting GPS would be expected to try to extend their campaign by seeking to evade detection and apprehension.[13] Approaches to do so could include tactics like delayed activation of jammers after attackers emplaced them, thus allowing escape (similar to use of timed explosives devices) or release of airborne jammers (e.g., on unguided balloons) that drift away from their point of release. However, such groups would likely become subject to increasing levels of investigative intensity as their campaign continued (similar to surging of resources to respond to other serial crimes).[14] Although emplacement or release of jammers would be different from traditional approaches to terrorism, traditional investigative approaches would still apply (e.g., broader collection of video and other evidence to identify individuals or vehicles near multiple emplacement or release sites). As a result, over time, the probability of the campaign being ended through law enforcement action would be expected to increase with each subsequent attack.

Because the attack itself involves transmitting in the GPS band at ground level, detection technologies could be very effective in making it possible to rapidly respond to individual incidents and in aiding in identifying the perpetrators involved. To the extent that these threats become more than theoretical, incentives for localized capability for more rapid detection of transmitters operating in the GPS band would increase.

Although the use of GPS jammers *could* be a tactic applied by nonstate groups, the choice to do so would be an unusual one based on the historical preferences of such groups for targets and tactics.[15] Past assessments of nonstate actor tactical choices in particular have demonstrated a level of operational conservatism among *most* such groups; they rely largely on proven, simpler, and more predictable approaches in attack planning versus unusual and more novel ones.[16] The factors driving this are multifaceted and include both resource and knowledge constraints that limit the addition of new technologies and tactics and learning effects that create incentives to rely on tried-and-true methods.[17] As a result, a group choosing to pursue PNT-focused approaches would be doing so at the expense of a variety of other proven options for producing similar destructive (e.g., direct attacks on infrastructures) or disruptive effects (e.g.,

---

[13] In a National Risk Assessment (DHS, 2011), elicited experts predicted that most jamming scenarios were estimated to produce intermediate levels of disruption for about a week or less (in part determined by assumptions regarding the time it would take to locate jamming sources), except for use of many lower power jammers where attackers could extend their campaign and where effects were thought to be maintainable for more than a month. Use of spoofing was thought to potentially produce more consequential effects (potentially by directly causing damage to infrastructure systems) for extended periods.

[14] See, for example, analyses of arrest probability for serial offenders in Lammers, Bernasco, and Elffers (2012).

[15] See discussion in Jackson and Frelinger (2008) and Horowitz, Perkoski, and Potter (2018).

[16] For example, Jackson and Frelinger (2009); van Dongen (2014); and Knight, Kean, and Murphy (2017).

[17] Over the last decade, a significant amount of research effort has focused on innovation and learning efforts in terrorist groups and how their incentives and constraints affect their ability to utilize novel attack methods. Examples from among this body of scholarship include Jackson et al. (2005a, 2005b); P. Gill et al. (2013); Kettle and Mumford (2017); and Gill (2017).

repeated small events, use of threats and hoaxes) that are comparable to or even lower cost than PNT attacks. That a group might choose to do so would not be unprecedented. However, it would be a deviation from general behavior, and this must be a factor in any decisionmakers consideration of its likelihood and associated risk.

While much of the focus of concern with respect to nonstate actors is the use of GPS jamming technology, their potential to attack other components of the PNT ecosystem must also be considered given that nonstate actors are still strategic actors. If such a group is interested in targeting GPS, they logically might also pose a threat to other complementary PNT systems that either already exist or might be added to the PNT ecosystem. For terrestrially based components that have concentrated sites (e.g., a few large transmission towers for an alternative PNT signals), such sites would be subject to attack by a strategic actor seeking to deny PNT. Nonstate terrorist groups have a long history of attacking such tower sites. Since 2000, there have been more than 500 attacks of various types on utility and telecommunications towers worldwide (e.g., electricity towers but also transmission and cell phone towers).[18] One of the events occurred in the United States in 2009 where two transmission towers at a radio station in the state of Washington were destroyed using a bulldozer in an attack claimed by the Earth Liberation Front.[19] Such physical operations would require a group to divide their resources and take additional operational risk, but it would likely be within the capabilities of many nonstate groups (Jackson and Frelinger, 2009). Other elements of the ecosystem (e.g., Wi-Fi signals, cellular signals, and local positioning systems) could also be vulnerable to jamming-type attacks, although the difficulty of the jamming operation would significantly increase because of the multiple frequencies involved and often their higher power.

The criminal use of GPS jammers can also be associated with concealing the location of stolen cargo or with concealing the locations of persons or vehicles who have either an overt or hidden GPS tracking device on them. Such uses of jamming are more akin to those captured in in the negligence category discussed above because the technology is being used to achieve a narrow purpose and because having a consequential effect on PNT more broadly is not the goal. Unlike the nonstate actor seeking to jam a large area or achieve particular effects on critical targets to achieve strategic objectives, criminal actors have a much narrower set of operational goals, ones usually focused on denying the ability of law enforcement or others to track them. The effects of interest to us in this study are on the collateral impacts of the use of the jamming systems that impact nontarget users. Criminals using focused attacks on GPS users, such as against a construction site to extort money by interfering with business, are

---

[18] START Global Terrorism Database, incidents from 2000–2016 on target types "utilities" and "telecommunications," where the target description included the word "tower."

[19] START Global Terrorism Database, Event ID 200909040003.

treated as a focused small-scale attacker and would have negligible effects at the scale of effects examined in this study.

A Specific Case: Cyberattacks Against GPS

As a technical system reliant on information technology, hacking and cyberattacks could be a route to disrupting or corrupting the function of GPS or other PNT systems. Incidents of unintentional software-based disruption have occurred—notably the 2016 data upload problem that resulted in a timing error across the GPS system,[20] an incident that demonstrated the potential for disruption or damage caused by intentional adversary action.

Although theoretically possible, cyberattacks on the GPS system itself appear to be quite difficult to stage in practice. First, as a military system, GPS is both better protected and more isolated from cyber intrusions than most systems; the most relevant of such systems for this analysis are the varied commercial and other proposed backups or alternative systems that could supplement GPS. If GPS could be so attacked by way of cyber means, logical consistency requires accepting the potential for similar attacks on those other systems. Second, to be effective, such an attack could not be limited to GPS. As discussed above for acts of war and PNT terrorism, the availability of alternative GNSS signals and the increasing integration of technology for their utilization in even commercial-level systems complicate this scenario. As a result, to have the most significant effect, attackers would also need to disrupt the alternative GNSS systems that are protected in a similar manner to GPS. Consequently, we find such attacks to be the province of nation-states, not terrorist or other small groups; and the nation-states themselves both depend on and own some of the alternative GNSS systems.

The practical difficulties of a cyberattack still remain. If the attacker is not itself the Russian state, it would have to attack GLONASS and GPS. Soon, it would also have to attack the European Galileo and the Chinese BeiDou as well as any additional British system if a British GNSS is indeed built (Clark, 2018). The difficulty of a multi-pronged cyberattack against multiple different nation-states appears unattractive compared with a more direct and straightforward cyberattack against alternative targets that could themselves produce national-level effects on the United States (for example, Supervisory Control and Data Acquisition [SCADA] systems of the U.S. power grid), which are thought to have their own cyber vulnerabilities.

Nonetheless, it is in principle possible for a cyberattack to disrupt all the relevant GNSS and result in a long hiatus of PNT services from space. Such attacks would seem to require the resources of an advanced nation-state, such as Russia and China. But there is a problem with imagining such a cyberattack as a rationale for any backup: The

---

[20] Described, for example, in U.K. Government Office for Science (2018).

backup would not be more secure than GPS, much less more secure than the combination of multiple GNSS constellations, especially for a terrestrial backup system that would then be more vulnerable to physical disruption or destruction. We will return to this point in Chapter Seven.

Summary

In considering the threats and vulnerabilities of the GPS system and its alternative PNT system, GPS itself, as a system, is generally very robust. Plausible potential threats from human actions and natural hazards are of very limited duration—days at most—and most threats are also quite limited in areal coverage. Moreover, essentially all the applications of GPS have existing backups or workarounds that limit or in many cases (timing for banks, timing for cellular towers, etc.) eliminate the problem.

Alternative PNT systems have their own set of risks, which differ from GPS. In general, however, centralized terrestrial systems are significantly more vulnerable than GPS to physical attacks. Distributed, diverse backups, much as we now have, offer the most resilient backups.

CHAPTER FOUR

Conduct an Economic Analysis of the Damages Resulting from Varied Threats to PNT

As shown in Chapter Three, GPS and PNT systems more broadly are subject to a range of threats and natural or accidental operational degradation or interruption. But assessing whether investments should be made to increase PNT resilience—or, specifically, to back up GPS in particular—requires first estimating the damages associated with disruptive incidents. In this chapter, we address the second part of the analysis: *conduct an economic analysis of the damages resulting from the varied threats to PNT.* This part of the analysis addresses the following questions from Table 1.1: How much damage is likely, by sector of the economy and aggregated for each threat in Chapter Three? And importantly, what are the estimates of the effects, given existing backups and workarounds?

Our bottom line answer is that when estimates of the cost of GPS disruption or loss include realistic adaptation options and existing complementary technologies (backups and workarounds), the total of the relevant, preventable damages across sectors is, in context, modest. Even for nationwide disruptions, as with a large CME, the preventable damages are only in the hundreds of millions of dollars per day; local outages, as from jamming, are a great deal smaller. While this estimate for the nationwide total is hardly small, it is quite modest in its proper context—when compared with the costs of alternatives that could mitigate these damages. We turn to that comparison, in part quantitatively and in part qualitatively, in Chapter Six after introducing the alternatives and more quantitatively in Appendix G, relying there on additional, proprietary information.

In the remainder of this chapter, we discuss our approach to conducting the economic analysis. The chapter ends with a discussion of the damages expected from both a limited, regional disruption as well as the much more consequential damages from the most stressing, but still plausible disruption we found—an extreme CME.

Approach for Conducting the Economic Analysis

In making estimates of the damages associated with the disruption of GPS and/or other GNSS, we addressed several key issues to reduce unrealistic overestimation or underestimation.

- First, analysis reflected the range of existing capability available in the PNT ecosystem because a variety of systems and complementary technologies are already broadly deployed or will be in the near future. In some cases, these other ecosystem components already provide considerable resilience to national PNT capability for some applications. Examining threats to GPS in isolation and neglecting the potential resilience of the PNT ecosystem to help compensate for disruption or outage would result in unrealistically high estimates of potential costs.
- Second, just as complementary technical systems provide resilience to disruption of GPS or GNSS, the legacy of data produced from past usage can also bridge periods of disruption for some applications. While GPS may be critical for creating precision maps of urban areas for navigation or farmland for crop planning, maps "on the shelf" retain their value until enough changes in the natural or built environment force their updating. Neglecting the value of the data legacy that has resulted from the past use of PNT will skew the estimates of the costs associated with disruption higher.
- Third, estimates of the damages from GPS disruption (and, therefore, the benefits of backup or resilience investments) must consider the human and organizational ability to adapt and compensate for disruptions in PNT availability. While there may not be adaptation options available for constant, real-time uses of PNT, in other cases delaying tasks that needed PNT, falling back to other ways to perform those tasks, and relying on other options are possible—usually with some penalty in efficiency. Although some have suggested that people and organizations are becoming increasingly dependent on PNT—for example, losing the ability to navigate without turn-by-turn directions while driving—assuming that such dependency is broadly the case will unrealistically bias cost estimates upward. In reality, day-to-day variability in conditions means that individuals and organizations must adapt to many different events, and such behaviors realistically would help lessen the impact of PNT disruption or corruption.

Estimating Damages Due to a Nationwide Disruption of GNSS

We draw on published estimates of the losses to the U.S. economy that would result from loss of GPS services, especially from a current report by O'Connor et al. (2019)

and by earlier reports by Leveson (2015a; 2015b), and by Pham (2011).[1] We extrapolated the older numbers from Leveson (2015b) and Pham (2011) to a range of possible values for the present day, making allowance not only for the growth of the economy but also for the increased adoption of and increased dependence on GPS (thus, the high range of this extrapolation may be much higher than Leveson's or Pham's original numbers—see Appendix A for details). To a lesser extent we also draw on Sadlier et al. (2017), which was a study of potential losses to the U.K. economy during a GNSS disruption. Finally, for some sectors we made our own independent estimates of the economic losses that would result from a GNSS outage.

One reason it is necessary to draw from multiple studies is that each study had certain sectors it focused on and others it neglected, as shown in Table 4.1. Also, is instructive to consider the areas where multiple studies roughly agree and the areas in which different studies arrived at very different estimates. Each estimate rests on assumptions of one kind or another, and it is fair to say that any single estimate has a high level of uncertainty.

Data Source Caveats

The economic impact numbers from different studies must be interpreted with some care.

In many cases, other authors have estimated that the availability of GPS provides a certain economic benefit to a given sector that would correspondingly be lost in the event of a GPS outage. In other cases, however, the "benefit" from GPS is calculated as the net benefit over a hypothetical alternative that could have been built—for example, Leveson made this kind of counterfactual calculation for timing applications in the communications, finance, and electrical sectors, though not generally for other applications. Therefore, Leveson's benefit estimates for timing applications cannot be interpreted as the economic loss that would result from a surprise GPS outage, when there would be no opportunity to build the hypothetical timing alternative.

For the other sectors, Leveson based his economic gain estimates on either (1) estimated productivity differences between activities that use or do not use GNSS services or (2) willingness-to-pay estimates for GNSS services. For these sectors we considered the economic losses that would result if those activities that lost GNSS would revert to productivity equal to those that were not using it anyway and if willingness to pay for services indeed represented the loss associated with loss of access to the services. We note that plausible arguments can also be made either for larger impacts, due to,

---

[1]   Leveson (2015b) and Pham (2011) themselves draw on many other studies of the economic benefits of GPS. Some more examples of such studies are ACIL Allen Consulting (2013); Carroll and Montgomery (2008); European Global Navigation Satellite Systems Agency (2016, 2017); Grisso, Alley, and Groover (2005); John A. Volpe National Transportation Systems Center (2009); John Deere (2018); Martin et al. (2005); Oxera Consulting (2013); Schimmelpfennig (2016); Technical Strategy Leadership Group (2012); U.K. Government Office for Science (2018); and Young, Rogawski, and Verhulst (2016).

say, loss of knowledge about how to carry out activities without GNSS, or to smaller impacts, due to, say, postponement of activities until GNSS service is restored.[2] We also modified the estimates from Leveson to reflect more plausible finite elasticities of supply and demand—details of this modification are explained in Appendix A.

Having estimated the economic benefits of GPS for several commercial sectors, Pham (2011) made the assumption that the economic benefits for all *other* commercial sectors were proportional to the relative spending of those sectors on GPS equipment over the previous six years (2005–2010). In Table 4.1, we show the results of carrying this logic one step further and assuming that the economic benefits for those other sectors were proportional to relative GPS spending, not merely in the aggregate but sector-by-sector. We recognize that this assumption is not ironclad, and the ratio of benefits to spending might actually vary from sector to sector. Nevertheless, bearing in mind that we expand Pham's numbers into a range of estimates, it is interesting to consider that GPS benefits to a sector might be roughly commensurate with GPS investment in that sector.

O'Connor et al. (2019) and Sadlier et al. (2017) estimated in some cases that the damages from a GNSS outage would *far* exceed the mere loss of GNSS benefits—that instead of being able to revert to a pre-GNSS level of efficiency, some sectors would be hardly able to function for a significant period if deprived of GNSS. We consider these worrisome estimates to be in the range of possibilities as well. However, the historical evidence for these pessimistic "shock" scenarios is not strong. When actual GNSS denial events have occurred in the past, they have not caused such dramatic economic shutdown. We will review this historical evidence at the end of this chapter.

Because our analyses examine potentially GPS-dependent sectors individually, our estimates of costs are based on national economic data for individual sectors. As a result, unlike more elaborate economic models that explicitly capture the interconnection of different industries, these estimates do not capture the indirect effects that disruption of one sector might have on other industries. Extrapolating to try to capture such indirect effects was beyond the scope of our work, but the values we report could be thought of as having multipliers that would increase the total costs. For example, national-level input-output analysis results from the U.S. Bureau of Economic Analysis show multipliers for different industries in different regions of the country. Publicly published examples of these multipliers for some regions range from just over 1.0 to just over 2.0—that is, a dollar of activity in a sector might produce more than an addi-

---

[2]   These opposite arguments are both more plausible for the first few hours or days of a GNSS outage than they are for a prolonged outage. On the one hand, it is plausible that a short disruption would be "just a day off" and activities would merely be rescheduled with almost zero real damage to the economy; but this argument is less applicable to long GNSS disruptions. Likewise, it is somewhat believable that some workers could be confused and literally stop work at the outset of a GNSS outage; but it is less believable that these workers would remain inactive and not find ways to proceed without GNSS if the outage continued.

tional dollar of additional output in other sectors.[3] However, naïvely applying multipliers to the values reported should also be done cautiously because in many industries, some of the costs of GPS disruption could be recouped through rescheduling activity or other resilience measures. This is consistent with literature on resilience approaches limiting the costs associated with natural disasters or other disruptions of economic activity (Rose et al., 2007). Here, we did not use any multiplier. A multiplier between 1.0 and 2.0 would not fundamentally change our analysis.

Additionally, large-scale, multiday events that could disrupt GPS around the world, such as a CME, disrupt multiple sectors at once and so introduce the potential for nonlinearities in estimates. These are fundamentally uncertain in nature and, in the case of a CME, might be further compounded by effects on the power grid, though the grid operators have taken steps to mitigate their vulnerabilities to a CME (GAO, 2018). For this analysis, it is important to note the *lack* of any evidence for large nonlinearities, either in the sector-specific analyses or in the historical examples summarized at the end of this chapter. Given the absence of any evidence for such effects, we do not include them.

The range of economic damage estimates from different sources is summarized in Table 4.1. We discuss each sector in detail below and explain which estimate we will use.

Estimating Damages from Regional GNSS Jamming

In addition to considering nationwide outages, we also considered a number of scenarios in which localized jamming was directed at achieving a particular effect. This is a more achievable and thus a more realistic scenario for a malicious attack than continent-wide jamming. Each scenario focuses on a critical sector and a major city or area where there was a significant amount of economic activity in the sector based on national economic data.[4] Using map data for the area, the distribution of the specific

---

[3]   For example, the Bureau of Economic Analysis ([BEA] 2010) shows output multipliers for a region in New Jersey and Delaware where the cross-sector effects ranged from a low of 1 to a high of 2.13.

[4]   One major source of data for the gross domestic product (GDP) data cited in the following sections is BEA (2018b). Because this data only includes data on Metropolitan Statistical Area (MSA) contribution to the nation's GDP, data for countrywide sector contributions was taken from BEA (2018b). Value added is described by the BEA as: "Value added is a measure of output after accounting for the intermediate inputs used in production. As such, it is also a measure of an industry's contribution to GDP" (Streitwieser, 2011, p. 7). Another important source of data for this analysis was the Census Bureau's 2012 Economic Census as well as the Department of Agriculture's (USDA) corresponding 2012 Census of Agriculture. This data was required to supplement the BEA data above, primarily because the BEA only reported data at the sectorwide level while the Economic & Agriculture census report data at the subsector level or below. Data for some businesses for which the value of GPS benefits has been described as in the high billions of dollars, such as Surveying, can only be tracked down at the North American Industry Classification System (NAICS) industry level. Data on the subsector, Securities, Commodity Contracts, and Other Financial Investments and Related Activities, an area where GPS's

Table 4.1
Estimates of Daily Cost of Losing GPS Services Nationwide (millions of then-year dollars)

Sector	O'Connor (2017$)	Pham (2011$)		Leveson (2013$)	Extrapolated Range from Leveson and Pham (2018$)			HSOAC (2018$)
					Low	Medium	High	
Consumer location-based services	95	N/E		36	44	128	254	N/E
Commercial road transport	138	28	28	33	20	61	123	N/E
Emergency services	N/E[a]	N/E		N/E	N/E	N/E	N/E	11–72
Agriculture	503[b]/41[c]	553[b]/54[c]	553[b]/54[c]	381[b]/38[c]	220[b]/22[c]	340[b]/34[c]	500[b]/49[c]	N/E
Construction	N/E	25[d]	10[e]	14	8	24	43	345
Surveying	11[f]		15[e,g]	32[g]	14[g]	15[g]	17[g]	7[f]/11[g]
Aviation	N/E	77[d,e]	7[e]	0.4	4[e]	6[e]	9[e]	N/E
Maritime	123[h]		61[e]	0.5	36[e]	58[e]	85[e]	N/E
Railway	N/E		0.1[e]	0.1	0.1[e]	0.1[e]	0.2[e]	N/E
Timing (telecom)	40–456[i]							0
Timing (electric grid)	9		9[e]	0.1[j]	6[e]	9[e]	13[e]	N/E
Timing (finance)	0							9
People tracking	N/E		0.6[e]	N/E	0.3[e]	0.6[e]	0.8[e]	N/E
Port operations	224[k]	N/E		N/E	N/E	N/E	N/E	0–880
Mining	32	N/E		N/E	N/E	N/E	N/E	N/E
Oil and gas	51	N/E		N/E	N/E	N/E	N/E	N/E

precision timing functions might be of extreme salience, would likewise prove relatively intractable unless one uses this type of data source. According to the NAICS, various sectors are defined by various levels of aggregation, the highest of which is the "sector" level (e.g., Mining, Quarrying, and Oil and Gas Extraction), followed by the "subsector" level (e.g., Mining [except oil and gas]), "Industry Group" (e.g., Metal Ore Mining), "NAICS Industry" (e.g., Gold Ore and Silver Ore Mining), and finally, the "national industry" (e.g., Gold Ore Mining). For more information on how the sectors are classified, interested readers are encouraged to read Office of Management and Budget ([OMB] 2018). The NAICS system is under constant revision, and because the Economic & Agricultural censuses were taken in 2012, the 2012 NAICS classification system was used to ensure compatibility. One noteworthy difference between the Economic & Agricultural census data and the BEA's data is that the former breaks out receipts (or revenue) and payroll separately and does not seek to aggregate them into GDP. The USDA's Agricultural Census describes the dollar value it attributes as the "Market Value" of the goods produced. Thus, the two figures are not directly comparable, and readers should avoid making comparisons between Receipts data and GDP data. While Receipts are one portion of GDP, they do not tell the whole story, and given the time and resource constraints of the project, reliable estimates of value added GDP were unable to be derived from the available data. Finally, the Census Bureau, and USDA, do not report data at this level aggregated by MSA, but rather by county. Because MSA's conform to county, or equivalent entity, boundaries, a crosswalk was compiled using the Federal Information Processing Standard Publication county codes, to assemble as comparable an entity as possible.

Table 4.1—Continued

SOURCES: O'Connor et al., *Economic Benefits of the Global Positioning System (GPS): Final Report*, Research Triangle Park, NC: RTI International, RTI Report No. 0215471, June 2019; Pham, *The Economic Benefits of Commercial GPS Use in the U.S. and the Costs of Potential Disruption*, New York: NDP Consulting, June 2011; Irv Leveson, "Recognizing GPS Contributions," presentation to the National Space-Based Positioning, Navigation, and Timing Advisory Board, Annapolis, Md., May 7, 2013.

NOTE: N/E = not estimated. Note also that although these numbers are often presented with up to three significant figures, that is only to ease any numerical comparison across this document; the actual precision of all these estimates is much less.

[a] O'Connor et al. (2019) decided not to quantify the benefit of lives saved by emergency responders but did assess consumers' willingness to pay for more accurate emergency call location as a part of location-based services.

[b] Daily cost for an outage occurring at a critical period of agricultural activity (planting season).

[c] An "average" expected daily cost for an outage occurring at a random time of year.

[d] Pham (2011) combined these sectors when estimating GPS benefits.

[e] This estimate rests on an assumption that GPS benefits are roughly proportional to GPS spending across sectors and based on Pham's (2011) data from 2005–2010.

[f] Excludes surveying for the mining and oil and gas sectors.

[g] Includes surveying for the mining and oil and gas sectors.

[h] Includes recreational boating (the dominant component), commercial fishing, navigation in seaways, and towing. Does not include cargo operations in ports, which are listed separately.

[i] In the O'Connor et al. (2019) model, costs would start in a range from $0 to $108 million on the first day and then escalate, to a range of $290 to $571 million per day after 30 days; we have listed their estimate labeled "average."

[j] Leveson (2013) estimated very small timing benefits from GPS, on the assumption that an alternative global time system could be built, but these numbers are not meaningful for a sudden GPS outage with no global alternative in place.

[k] Average daily impact from a 30-day outage. In the O'Connor et al. (2019) model, daily impact would initially be much less, but would grow larger with the duration of the outage.

industry being examined, and footprints of jammers (Table 3.1), we analyzed a range of jamming scenarios that took into account the local economic geography, geographic features and built environment, and other factors to estimate outcomes.

Table 4.2 shows the estimated costs resulting from these regional jamming scenarios. The scenarios are described in more detail in the following text of this chapter.

We now consider individual economic sectors in some more detail.

Consumer Location-Based Services

GPS is part of many activities that directly touch the lives of consumers. The modern smartphone is by far the most common GPS receiver that individuals interact with on a daily basis. This technology is part of a complex array of systems that provide consumers with many services. These include time-saving applications that provide route information, ride-hailing services, lifestyle applications (e.g., dining, dating, fitness, and shopping), and even life-saving applications to call for aid in an emergency. This vast web of applications all benefit from the existence of GPS as part of the smart-

Table 4.2
Estimated Daily Cost of Losing GPS Services in Targeted Regional Jamming Scenarios (millions of 2018 dollars)

Targeted Sector	Scenario Location	Estimated Daily Cost (Millions of 2018 Dollars)
Consumer location-based services	Chicago	3.2
Agriculture (wheat)	Kansas	0.4–1.5
Construction	Houston	5
Surveying	Atlanta	0.05
Finance	Los Angeles	1
All sectors equally targeted	Los Angeles–Long Beach	10–11

phone's location services that are used by all the applications. The location services on many phones also make use of several GNSS systems (including, but not limited to GPS), cellular tower locations, and Wi-Fi sources as well as Bluetooth fixes from beacons, onboard compasses, accelerometers, and barometers. Indeed, while many consumers assume GPS is the sole source of positioning information for their phones, in reality a mix of onboard and off-board data sources are used.

Phones are an undeniable feature of modern life, and GPS is so ubiquitous that finding a phone without some GPS functionality is extremely difficult. However, despite this ubiquity, the loss of GPS to consumers would lead to a loss of efficiency and some inconvenience rather than total disruption of day-to-day activities. According to the Federal Highway Administration's (2017) *National Household Travel Survey*, the destinations involved in most consumer trips are to a small set of stable sites that are accessed by well-known routes, such as the path between work and home.[5] For routine transport functions, especially in familiar environments, studies have suggested route choices may not vary much (Zhu and Levinson, 2012). We think this casts considerable doubt on the assumption in Sadlier et al. (2017) that a GNSS outage would lead to an 18-percent delay for all road users. The great majority of drivers at any given time are on familiar routes and would not be delayed.

Given this proliferation of the technology, what is the best method to estimate the value consumers place on having GPS in their phones, and, conversely, what value is denied to those same consumers if an adversary jammed GPS for a brief period? One method of estimating the nonmarket value of a good is known as contingent valuation. Contingent valuation methods typically depend on asking consumers either

---

[5]   The purposes for car trips include going home (34.3 percent), work (16.6 percent), school/daycare/religious activity (3.1 percent), medical/dental services (1.5 percent), shopping/errands (19.5 percent), social/recreational (8.4 percent), transporting someone else (8.5 percent), meals (6.7 percent), and other purposes (1.4 percent) (Federal Highway Administration, 2017).

their *willingness to pay* to acquire a good, or their *willingness to accept* going without it (Hanemann, 1991).

O'Connor et al. (2019) surveyed location-based-service users in 2017 and found a mean willingness to pay of $8.11 per month ($8.29 in 2018 dollars) for turn-by-turn navigation services specifically and $13.17 per month ($13.46 in 2018 dollars) for all location-based services.[6] (There was, however, large uncertainty in the estimates.)

For historical comparison, Patel (2009) found that consumers in 2008 were willing to pay $3.80 ($4.45 in 2018 dollars) for maps and directions. Leveson (2015b) guessed that navigation and maps were worth $5 per month in 2013 ($5.42 in 2018 dollars) to mobile phone users and worth $12.50 per month ($13.56 in 2018 dollars) to users of automotive navigation systems. It is plausible that average willingness to pay has increased over the years as the quality and functionality of location-based services have increased.

Regional GPS Jamming Vignette: Chicago

To apply this estimate to a realistic jamming scenario, we selected the Chicago, Illinois, MSA. Chicago is the third most populous MSA in the nation, and the population distribution in the city includes very high population density areas where jamming would affect many people and therefore presumably many smartphone users (Figure 4.1).

Although urban environments create the potential for GPS jamming to affect many people at once, the built environment in major cities is a complex one for transmitting radio signals; the reception of satellite signals is affected by the "urban canyons" formed on streets surrounded by high buildings. That built environment would also reduce the efficacy of jamming because lines of sight are frequently interrupted by tall steel buildings (Figure 4.2).

To estimate the potential costs associated with a GPS jamming incident in Chicago, we consider the number of smartphones in the area and the willingness of users to pay for maps and location services on those devices, as estimated by O'Conner et al. (2019).

The Chicago-Naperville-Elgin MSA was estimated to have 9.5 million people in 2017 (U.S. Census Bureau, 2018). The World Bank (2018) reports that in 2017, there were approximately 120 mobile cellular subscriptions per 100 people in the United States, and so we infer there are about 11.4 million cell phones in the Chicago MSA. The Pew Research Center (2019) estimates that in 2019, 96 percent of U.S. adults owned a cell phone and 81 percent of adults owned a smartphone. From this we infer

---

[6] O'Connor et al. (2019) also found a mean willingness to accept $98.35 to do without turn-by-turn navigation for a month and $150.97 to do without all location-based services for a month. It is typical that willingness-to-accept estimates are significantly larger than willingness-to-pay estimates.

Figure 4.1
Population Density of Chicago

Population density (people/square mile)
■ 25,000 or more
■ 12,500 to 24,999
■ 7,500 to 12,499
■ 5,000 to 7,499
■ 3,500 to 4,999
■ 2,000 to 3,499
■ 1,000 to 1,999
■ 500 to 999
■ 250 to 499
Less than 250
■ Not applicable

SOURCE: U.S. Census Bureau, "Metropolitan and Micropolitan—Population Density by Census Tract: 2010," Census.gov, undated.
NOTE: Footprint assumes ten jammers placed on top of each 150-foot-plus structure to produce a 360-degree targeted arc to break signal lock for a majority of receivers. Footprint assumes a full jamming footprint of 95 km without geographic or horizon limitation.

that about 84 percent of cell phones are smartphones.[7] Allowing for some uncertainty in the above statistics, we estimate that there are on the order of nine million smart-phones in use in the MSA.

GPS jamming would not mean total loss of maps and directions on personal devices because (a) other signal sources such as cell towers and Wi-Fi would still provide some position fix to the device and also because (b) applications such as Google Maps and Yelp can still be used and still give some value even if the device has no position fix at all and the user is relying on her own knowledge of her position. O'Connor et al. (2019) made some allowance for this by supposing that of the $13.17 per month

---

[7] This inference might not be entirely correct—for example, it could be that children's cell phones are either more likely or less likely to be smartphones than adults' cell phones are, and it could be that the devices owned by users of multiple phones are either more likely or less likely to be smartphones.

Figure 4.2
Skyline of Chicago

SOURCE: Google Earth Pro, version 7.3, 2019.

in value, all the value attributable to navigation ($8.11 per month) would be lost but only half of the rest of the value would be lost for a net result of $10.64 per month lost. This may be slightly pessimistic since, as just stated, some form of navigation app use would still be possible. But on the other hand, we should consider that this monetary estimate of value to the consumer is somewhat conservative in that it reflects a willingness to pay and not the much higher monetary figure of willingness to accept doing without.

Applying the O'Connor et al. (2019) figure of willingness to pay $10.64 per month for maps and directions for each smartphone, we arrive at a daily lost value of $0.35 per device. If all smartphones in the Chicago-Naperville-Elgin MSA were denied location-based services for a day, the resulting loss to consumers—based on these figures—would be on the order of $3.2 million dollars per day in 2018 dollars.

As the skyline of Chicago shown in Figure 4.2 suggests, positioning a reasonable number of jammers that would cover the full population of the MSA would be difficult at best, with even airborne jammers affected by shadowing given the number of skyscrapers in the city center. As a result, in a real-world incident, the percentage of the affected area would be reduced.

Nationwide GNSS Outage

O'Connor et al. (2019) estimated the loss due to denied consumer location-based services at $95 million per day. This falls in the center of the range extrapolated from

Leveson (2015a, 2015b) ($44 million per day [low estimate], to $128 million per day [medium estimate], to $254 million per day [high estimate]), so we will adopt the O'Connor estimate, inflated to $97 million per day in 2018 dollars.

Commercial Road Transport

Pham (2011) and Leveson (2013) both focused on the value of GPS to commercial road transport, and our extrapolation from their figures gives a range of estimates from $20 million per day (low estimate), to $61 million per day (medium estimate), to $123 million per day (high estimate). O'Connor et al. (2019) estimated the value of GPS to the "telematics" sector, defined as the use of "in-vehicle equipment to remotely monitor vehicles for a variety of purposes," which largely overlaps with the commercial road transport sector. They estimated telematics losses due to a nationwide GNSS outage at $91 million per day (low estimate), to $138 million per day (medium estimate) to $184 million per day (high estimate). The majority of these vehicle fleet efficiency losses would be due to increased labor costs, with a smaller share of losses due to increased fuel consumption, increased vehicle repair and maintenance costs, and increased environmental costs.

We think these estimates probably overstate the true costs of a GNSS outage because they tend to ascribe all efficiency benefits of computerized vehicle fleet management to GNSS. During a total outage of GNSS, however, it would still be possible to plan efficient travel routes with digital map services such as Google Maps. It would still be possible to choose routes based on real-time traffic data.[8] Without turn-by-turn alerts from a GNSS system, there would be an increased cognitive burden on drivers, but they could likely plan and follow the same routes through a combination of being alert to street signs, consulting digital maps, and if necessary, determining rough location through cell phone triangulation. In short, not all of the benefits of digital navigation would go away during a temporary GNSS outage. However, given available information, it is harder to guess what fraction of benefits would go away.

In conversation with the Federal Highway Administration (FHWA) and National Highway Traffic Safety Administration (NHTSA), no direct dependencies on GPS were apparent for the infrastructure and safety requirements of the highway system.[9] With regard to truck fleets, it was noted that teamsters operated without GPS for

---

[8] Real-time traffic data from fixed road-based sensors, generally on highways, would be unaffected by a GNSS outage. Traffic model predictions based on accumulated historical data would also be unaffected. Real-time traffic data on local streets is derived primarily from tracking cell phone locations, so if the accuracy of cell phone locations was degraded by a loss of GNSS, then that element of traffic data would also be degraded to some unknown extent—the degradation might not be severe, since only an averaged measurement of traffic motion is needed.

[9] Conversation with NHTSA and FHWA, February 4, 2019.

decades. No autonomous vehicles are GPS dependent. GPS may be a part of an over-all autonomous or auto-pilot system but is not on its own capable of keeping a vehicle safely on the road. Crash avoidance systems do not use these systems. At no point is GPS a matter of safety of life.

For our aggregation later for nationwide losses, we adopt the middle figure from O'Connor et al. (2019) as a reasonable estimate, inflated to $141 million per day in 2018 dollars.

Hazardous Material Transport

The John A. Volpe Transportation Systems Center (2001) noted that the main criti-cal applications for GPS in road transport are in Intelligent Transportation Systems Hazardous Materials (hazmat) transport and for first responders in emergencies. We will discuss hazmat transport here and discuss emergency services as a separate sector.

The costs associated with hazmat accidents on roads are routinely tracked and compiled by the DoT. A 2007 study stated that transportation by truck accounted for over 90 percent of individual hazmat shipments, representing approximately 40 per-cent of the tonnage (Erkut, Tjandra, and Verter, 2007). According to DoT statistics, hazardous materials accidents inflicted damages of almost $600 million between 2008 and 2017, with an average cost per year of $60 million (DoT Hazmat Intelligence Portal, 2018a). With approximately 143,000 incidents over that period, the damage per incident is about $4,200. Incidents involving hazardous materials also accounted for just over 100 fatalities and around 1,500 injuries over the same period. However, it is unclear to what degree PNT disruption or increased resilience might result in an increase or decrease of this total. From an economic point of view, these costs would be very small compared with even low estimates of the efficiency losses described above.

Emergency Services

To estimate the value of GNSS to the emergency services sector, we adapt the calcula-tions of Sadlier et al. (2017), Lumbreras and Machado (2014), and Jaldell, Lebnak, and Amornpetchsathaporn (2014).

Jaldell, Lebnak, and Amornpetchsathaporn (2004) estimated the cost of a delay in ambulance response time in Sweden at SEK17,244 per minute or $2,854 per minute in 2018 dollars.[10]

In a more recent study, Jaldell, Lebnak, and Amornpetchsathaporn (2014) com-pared ambulance response times with hospital outcomes based on data from the Thai emergency ambulance services and estimated that statistically, a one-minute improve-

---

[10] We have converted the figure in Swedish krona into U.S. dollars at the average 2003 exchange rate of 8.0787 SEK/USD (Board of Governors of the Federal Reserve System, 2019) and inflated to 2018 dollars using BEA data.

ment in an ambulance's response time had a 0.007-percent chance of saving a life and a 0.15-percent chance of mitigating a severe injury so that admission to hospitalization was not necessary. Jaldell, Lebnak, and Amornpetchsathaporn assumed that the economic value of mitigating a severe injury was 6 percent of the statistical value of a life.[11] Combining this with the $10 million value of a statistical life suggested by the DoT (2016),[12] it would follow that one-minute improvement in ambulance response time has a value of $1,600.

Sadlier et al. (2017) made use of the Jaldell, Lebnak, and Amornpetchsathaporn (2004) estimate, converted into U.K. pounds, to assess the costs of GNSS outage through delayed emergency response. Sadlier et al. (2017) further estimated that without GNSS, a lack of precise location data for cell phone emergency calls in a small percentage of the 31 million emergency calls in the United Kingdom each year could cause 1.1 million minutes of delay per year for an estimated annual cost of £1.9 billion ($2.5 billion) in the United Kingdom.

Lumbreras and Machado (2014) performed a similar calculation for the EU and estimated that 0.1 percent of the 320 million annual emergency calls in the European Union were serious calls lacking good location information, resulting in a total delay of 3.2 million minutes (low estimate) to 9.6 million minutes (high estimate). Lumbreras and Machado assessed the cost of this delay at €1300 per minute, based on the Jaldell, Lebnak, and Amornpetchsathaporn (2004) study—however, Lumbreras and Machado may not have adjusted their estimate for inflation.

National Emergency Number Association ([NENA] 2018) states that the United States has an estimated 240 million emergency calls per year—however, the date of this estimate is uncertain and it is likely a few years old. The National 911 Program (2017) reports that in 2016, a set of U.S. states and territories representing 81.3 percent of the U.S. population had 212 million emergency calls. Scaling this figure by population to the entire United States, we would infer that there were 259 million emergency calls in the United States in 2016, including two million emergency calls in Puerto Rico. Further extrapolating with overall population increase to 2018, we project 262 million emergency calls in the United States in 2018, including two million calls in Puerto Rico.

From these figures we form a low, medium, and high estimate of losses due to degraded emergency response time, as shown in Table 4.3. The largest source of uncertainty is in the amount of delay that would result from GNSS outage, followed by

---

[11] For comparison, the DoT (2016) suggests that a "moderate" injury be valued at 4.7 percent of a life, a "serious" injury at 10.5 percent of a life, and a "severe" injury at 26.6 percent of a life. However, Jaldell, Lebnak, and Amornpetchsathaporn's (2014) definition of a "severe injury" includes every injury that results in admission to hospital and likely spans all of these categories. Also, the mitigation of an injury so that admission to hospital is not necessary should probably have somewhat less value than avoiding the injury entirely.

[12] The DoT (2016) suggests valuing a statistical life at $9.6 million—adjusted for inflation, this is $10 million in 2018 dollars.

Table 4.3
Estimates of Nationwide Losses in 2018 Dollars Due to Increases in Emergency Response Time

Estimate		Low	Medium	High
Emergency Calls in the United States	Annually	240 million[a]	262 million[b]	262 million[b]
	Per Day	6,570	7,170	7,170
Extra Minutes Delay per 100 Calls		1[c]	2[d]	3.5[e]
Assumed Cost of One Minute Delay		$1,600[f]	$2,227[g]	$2,854[h]
Resulting Cost Per Day		$11 million	$32 million	$72 million

[a] Estimate from National Emergency Number Association, "9-1-1 Statistics," webpage, 2018.

[b] Extrapolated from National 911 Program, *2017 National 911 Progress Report*, Washington, D.C.: 911.gov, November 2017, and U.S. Census data.

[c] Low estimate from Cristina Lumbreras and Gary Machado, "112 Caller Location & GNSS," presentation at the European Emergency Number Association, Ref. Ares(2014)1665619, May 22, 2014.

[d] Medium estimate from Lumbreras and Machado (2014).

[e] Estimate from Greg Sadlier, Rasmus Flytkjaer, Farooq Sabri, and Daniel Herr, *The Economic Impact on the UK of a Disruption to GNSS*, London: London Economics, June 2017. The high estimate from Lumbreras and Machado (2014) was 3.

[f] Derived from Henrik Jaldell, Prachaksvich Lebnak, and Anurak Amornpetchsathaporn, "Time Is Money, but How Much? The Monetary Value of Response Time for Thai Ambulance Emergency Services," *Value in Health*, Vol. 17, No. 5, 2014, pp. 555–560, and U.S. Department of Transportation, *Revised Departmental Guidance 2016: Treatment of the Value of Preventing Fatalities and Injuries in Preparing Economic Analyses*, Washington, D.C.: U.S. Department of Transportation, August 8, 2016.

[g] Average of values derived from Jaldell et al. (2014); DoT (2016); and Henrik Jaldell, *Tidsfaktorns betydelse vid räddningsinsatser -en uppdatering av en samhällseknomisk studie*, Karldysf: Swedish Rescue Service Agency, Report P21-499/04, 2004.

[h] Estimate from Jaldell (2004).

uncertainty in the economic cost per minute that would result from that delay.[13] We will carry this range of estimates forward.

GPS Jamming in Conjunction with Other Attacks

One could also ask if there would be a synergistic effect of jamming GPS in order to interfere with the police and fire response to some other, different attack that was intended to cause casualties and induce panic. The key is that the only likely effect would be interfering with the location accuracy of cell phone users after an attack. But the majority of cell phone users and emergency responders would be familiar with their own locations at any given time—even setting aside cellular or Wi-Fi sources of posi-

---

[13] These calculations do not attempt to break down the distribution of delays between calls or assess nonlinear differences in the importance of delays. An observational study by Blackwell and Kaufman (2002) found that mortality risk was reduced for ambulance response times of less than five minutes compared with response times of over five minutes, but the mortality risk curve was generally flat for increases in response time over five minutes.

tioning—so from the attacker's point of view, this could at most sow a limited amount of confusion among a small fraction of the targeted population. It is still conceivable that a terrorist attacker would attempt this,[14] but for a similar level of effort and risk of exposure, the attacker would generally have better options for causing confusion—for example, by actually jamming emergency phone calls instead of merely distorting the location of the calls, and/or by placing false emergency calls or spreading false information.

Agriculture

One of the industries to benefit most from the widespread proliferation of GPS is the agricultural sector. The sector has benefited from the integration of GPS receivers into many of its activities, including for example farm planning and tractor guidance (National Coordination Office for Space-Based Positioning, Navigation, and Timing, 2018). Crops that can be harvested by machine enjoy benefits from GPS guidance while other crops such as grapes and strawberries are harvested by hand and GPS guidance is not used. Guiding equipment for pesticide, herbicide, and fertilizer application is another area where GPS is applied in farming, leading to savings by minimizing unnecessary application.[15]

Threats to GNSS in the Agriculture Sector

Farmland is often flat and wide open, offering little blockage to a GPS jammer. On the other hand, agricultural economic activity is spread across very wide areas, so malicious actors would be faced with the very difficult problem of needing to jam large regions to inflict meaningful economic damage.[16] The critical agricultural uses of GPS, however, are concentrated in a few weeks of the year during the planting and harvesting seasons; this is probably to a malicious attacker's benefit since the attack could be carried out at the worst possible time. The critical periods vary somewhat, however, by crop and by location.[17]

---

[14] One could draw an analogy to GNSS jamming during military operations, but the analogy is somewhat weak because military forces are often operating in unmarked or unfamiliar terrain and commonly direct fires by geographic coordinates—but police and emergency responders are generally operating in familiar territory and can usually make use of street directions.

[15] While area specific spraying for pests and weeds is a goal for some GPS usage, it is still more common for farmers to spray the entire field, recognizing that eggs and seeds often float and spread before the pests and weeds themselves appear (Adamchuk et al., 2008).

[16] The wide area dispersal of agricultural activity makes it difficult both for helpful actors to provide high-accuracy PNT to all users and for malicious actors to deny PNT to all users.

[17] For example, in Iowa peak planting season for corn is April 25–May 18; for soybeans, May 8–June 2. In South Dakota the peak planting season for corn is May 2–May 27; for soybeans, May 15–June 11 (USDA, 2010).

For natural threats due to space weather, these considerations are reversed. A space storm could affect GNSS availability to all agricultural users regardless of location. However, since any given space storm would presumably occur at a random time of year, there is a good chance it would *not* occur during a critical season, in which case agricultural losses would be minimal.

Regional Jamming Vignette: Kansas

To estimate the potential economic loss through the denial of GPS to agriculture, the focus must shift from metropolitan areas to more rural counties. As part of the country's agricultural heartland, Kansas, and specifically wheat farming, were chosen as the location and focus of this portion of the analysis. While corn is financially more important than wheat in aggregate, the wide dispersion of corn production made covering a significant fraction hard; in contrast, wheat appears to be more targetable. Figure 4.3 shows wheat production in Kansas, based on data provided by the U.S. Department of Agriculture.

Figure 4.3
Winter Wheat Production by County in Kansas, 2017

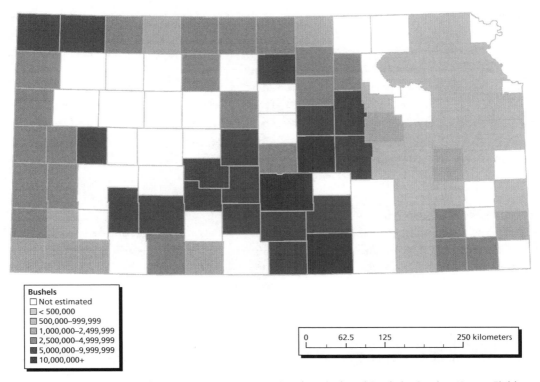

Bushels
☐ Not estimated
☐ < 500,000
☐ 500,000–999,999
▨ 1,000,000–2,499,999
▨ 2,500,000–4,999,999
■ 5,000,000–9,999,999
■ 10,000,000+

0   62.5   125   250 kilometers

SOURCE: U.S. Department of Agriculture, "USDA's National Agricultural Statistics Service, Kansas Field Office: County Estimates—Crops—Winter Wheat," webpage, 2017.

The USDA's Agricultural Census (USDA, 2016) reports that the market value of wheat products sold in Kansas was on the order of $2.5 billion in 2012 or $2.8 billion in 2018 dollars. Nearly all of the wheat grown in Kansas is winter wheat, with the most active planting between September 15 and October 20 and the most active harvesting between June 20 and July 5 (USDA, National Agricultural Statistics Service, 2010). A deliberate attacker might try to disrupt the active planting season of 36 days. Each day of disruption could potentially affect $77 million of planted wheat.

In thinking about an attacker trying to affect wheat production through GPS jamming, the key operational challenge is what fraction of the relevant land area under cultivation can be covered using a small number of jamming sites. For large areas, wide area coverage requires elevating the jammers, so we identified locations in Kansas where there were existing high radio or other masts (greater than 150 feet high) as candidate sites.

The amount of wheat production that could be disrupted from a single jammer site could possibly be as high as 28 percent of Kansas wheat production. The daily dollar value of potentially jammed wheat planting from this jamming site would then be 28 percent of $76 million or approximately $21 million. This represents an upper-end estimate because all wheat production in counties only partially covered by the jammer would not be disrupted and the range of the jammer in practice would be shorter because of horizon and topographic limitations.

Further, just because GPS was jammed in an area does not mean that a farm would shut down operations or halt planting or harvest. Equipment can still be operated manually, and analyses have calculated that the benefit in farming from use of GPS is an efficiency enhancement: GPS-driven planters allow farmers to avoid over-planting fields to save seed and fuel (Agriculture.com Staff, 2010). Estimated savings in 2009 put a GPS unit as providing a 2-to-7-percent cost savings per acre. Applying that percentage to the ranges of production at risk suggests a daily potential cost of GPS disruption of between $400,000 and $1.5 million, again in 2018 dollars.

Nationwide GNSS Outage

Our range of annual agricultural GNSS benefits extrapolated from Leveson (2013) and Pham (2011) is $7.9 billion (low estimate), to $12.4 billion (middle estimate), to $17.9 billion (high estimate). This benefit is concentrated in the planting and harvesting seasons for various crops, and these occur at somewhat different times in different places and for different crops. As a conservative bound, consistent-with-Kansas scenario, we might suppose there are 36 most critical days in the year for the most important crops, and denial of GNSS during those days might incur damages of $220 million per day (low estimate), to $340 million per day (middle estimate), to $500 million per day (high estimate).

The interviewees from O'Connor et al. (2019) noted that "The impact of a 30-day outage in the agricultural sector is highly dependent on the time in which it occurs. . . .

All agreed that the most damaging impacts would occur during planting season because farmers could be so delayed that they could potentially miss the planting window or plant at a suboptimal time, causing significant yield losses." It is important to recognize that "missing the planting window" requires that the disruption in GPS lasted throughout the 15-to-30 window for a particular crop at a particular location and latitude; shorter disruptions would allow some rushed planting to be supported by GPS. Assuming such a long disruption occurred, O'Connor et al. then estimated the losses through expert elicitation as "17% with a +/– 6% margin of error for revenue losses across corn, soybeans, wheat, rice, peanuts, and cotton." They estimated the potential damages to crops—most significantly, corn and soybeans—if there were a GNSS outage during the planting season, weighted by the estimated adoption rate for GPS systems in different crops, and added these damages together.[18] Their final estimate of the damages from lost revenue is $15.1 billion for a 30-day outage or $503 million per day. Thus, the high-end estimates are roughly consistent in the case where the GNSS outage occurred at a critical time for agriculture.

If the nationwide GNSS outage occurred during a random time of year, as presumably would be the case for a CME event, it would be a matter of chance whether agriculture was badly affected. If the outage occurred outside of major planting and harvesting seasons, the effect on agriculture would be small, perhaps zero. The "average" expected losses could be estimated as the annual benefit divided by 365.25—that is, $22 million per day (low estimate), $34 million per day (middle estimate), or $49 million per day (high estimate) from Leveson (2013) and Pham (2011). For O'Connor et al. (2019), the amount would be $41 million per day. We adopt the range in O'Connor et al. inflated to 2018 dollars ($42 million per day to $514 million per day).

Construction

Construction nationwide contributed about $580 billion to the national economy in 2012 (BEA, 2018a). The most significant portion of this sector from a GPS jamming perspective is site preparation because it involves earthmoving equipment that can make extensive use of automated machine guidance applications. GPS can also be used in site preparation planning and the compliance assessment of those plans. There are other applications of GPS in tracking and site management, but earthmoving applications appear to bring the most significant productivity gains and cost reductions.

---

[18] Adding together the potential damages for all crops is probably a minor overestimate of the worst case scenario since, as previously mentioned, the planting seasons for corn and soybeans only partly overlap and also vary by location, so a single outage is unlikely to disrupt the planting seasons for all crops everywhere. However, since the planting seasons for major crops do overlap to some degree, this error is not very large compared with all the other uncertainties in these calculations.

To a lesser extent, large civil engineering projects and line installation (both power and water/sewer) use GPS, but they also routinely require confirmation from other sources. These systems are one part of a family of construction techniques from multiple eras. At any given time, the construction industry has techniques and processes from ancient to cutting edge in application side by side. Crews can adapt to changes by applying hybrid techniques and field-expedient measures, permitting construction activity to continue even in the face of unexpected hiccups.

There are still critical paths in construction, and the construction managers and planners for such projects are responsible for guarding against failure along those critical paths. There are any number of things that can go wrong and delay a project: severe weather, raw material delivery and quality issues, labor availability and pay issues, technical failures of equipment, on-site accidents, fuel cost and availability, financial changes, local political and regulatory changes. GPS interference would not halt construction as a whole, but it could impact preparation, construction, and certification activities by impacting one of the subprocesses. A construction manager could account for GPS failure by recognizing the possibility of failure in the affected subprocesses, which are already vulnerable for other reasons. The novelty of GPS failure may catch some unaware and lead to inadequate response planning, but the ultimate impact of jamming is nothing new.

Regional Jamming Vignette: Houston

Not unexpectedly, the majority of construction activity occurs in large cities: MSAs accounted for roughly $565 billion, or 96 percent, of total construction activity where data was available. As a result, if attackers were seeking to impose costs through GPS disruption in the construction sector, they would almost certainly target a large city. As the MSA with the second largest share of construction activity, we selected Houston as a plausible case to anchor the upper bound of potential economic loss in the construction sector from GPS jamming. Houston's share of this total is roughly $31 billion in 2018 dollars (or about 4 percent of the national total in 2017) (BEA, 2018a)—this is $120 million per workday or $85 million per day for an outage that included both workdays and weekends.

To illustrate a jamming attack on construction in Houston, we selected high-powered but man-portable jammers (the middle row in Table 3.1 above) with an effective range of nine kilometers in the urban environment. There are thousands of permits for construction in the Houston MSA for the years 2017–2018. Some of these permits are for small changes to structures where disruption of GPS would be inconsequential. Others are for large building projects. A cursory sample of publicly available overhead imagery of the region turned up dozens of major construction sites both clustering in a core area and along major arteries. A subset of major sites were mapped to provide a real-world example target set to assess jamming effects.

We estimated that a realistic attack using two jammers could affect 28 percent of construction sites in the region. Of the $85 million per day ($120 million per work-day) total construction activity cited above, this would correspond to approximately $24 million per day ($33 million per workday). These would be the damages if (a) GPS jamming halted operations completely at all the covered sites and (b) none of the missed work could be made up. This is unrealistic, however, because not all major site construction activity even uses PNT and because significant construction operations routinely respond to disruptions in schedule from various sources (e.g., weather), balancing activities that are and are not affected by the changes in conditions.

Based on a number of construction cost estimation tools and examples of construction costs, excluding the initial surveying (which is addressed in the surveying section below), construction costs that are GPS intensive tend to include excavation, grading, and structural steel or structural concrete pours as well as regular checks. These disciplines can use GPS but can also be performed with the aid of a total base station or theodolite, and a bit of extra labor. GPS makes these more efficient but not necessarily more accurate. Assuming that the full cost of structural layer, grading, and excavation is due to GPS and is totally lost, then a random sampling of construction costs available online tended to converge around 15 percent. Taylor (2015) suggests a high end excavation and foundation cost of about 12 percent of the total price. Moselle (2017) estimates the cost of foundations at somewhere between 6 and 25 percent of the total cost of a building, including the labor, concrete, and steel.

Construction cost estimates typically include a contingency plan of at least 5 to 10 percent of cost coverage. The expectation is that changes in material prices, weather delays, overtime labor, and other factors may lead to some element of the budget over-running. Low-end examples of construction contingency costs can be found in military planning documentation (Unified Facilities Criteria, 2017).

Thus, based on available literature on construction, we estimate that 15 percent would be a conservatively high estimate of the portion of construction activities that are significantly GPS dependent and, therefore, a reasonable estimate of the fraction of jammed sites whose activities were vulnerable to disruption on any given day. This would result in a conservative estimate of $3.6 million per day ($5 million per work-day) in potential cost from a two-jammer-location operation.

These costs might be further reduced, however, by adjustments in schedule (i.e., "making up" GPS-related activities on a later date after the jammers have been located and neutralized). Most large construction jobs—the ones that would likely be using GPS at some point—have slack in their schedule to handle delays from such things as weather, accidents, missed deliveries, and so forth. However, the tasks for which GPS is normally used cannot be avoided indefinitely—if the excavation, grading, and structural work is not completed, the building will not be completed either. The actual damages from any outage would then depend on whether the disruption from GPS in some jobs, along with other delays, exceeded the built-in schedule allowance and so caused some increase in costs, as for later material. We lack the detailed information on schedules and delays to estimate this.

Nationwide GNSS Outage

A nationwide GPS outage would not likely be due to localized jamming but more likely would be due to a space storm, or some form of attack that disrupted the whole GPS constellation, and perhaps other GNSS constellations also. Therefore, our estimates above of how much metro activity could reasonably be covered by localized jammers are assumed not to apply. However, our estimate that 15 percent of construction activity is dependent on PNT would continue to apply.

Since all construction activity in the United States amounted to $840 billion in 2018 (BEA, 2019), disrupting 15 percent of this activity would cause damages of $345 million per day, assuming the work was evenly spread across 365 days. Seasonal effects would vary this amount as less construction occurs during the winter months in northern states. Nonetheless, we regard this as an unrealistic upper bound on the daily damages; it applies in the case where all excavation, grading, and structural work was actually brought to a halt by a GNSS outage, which we find to be unlikely. If the outage was not quickly resolved, we anticipate that construction crews would revert to pre-GPS methods of measuring positions and continue to work at pre-GPS levels of efficiency.

Other estimates are available, in this case drawing from both Leveson (2013) and Pham (2011). We derive from their work a range of nationwide economic losses between $8 million and $43 million per day, inflated to 2018 dollars. We will adopt a middle estimate, $24 million per day, in our overall total of potential damages.

Surveying

One industry that has benefited considerably from the proliferation of GPS has been the surveying industry. Surveying is a combination of office and fieldwork where fieldwork is most affected by the capabilities provided by GPS. However, the majority of surveying is not fieldwork, and the majority of fieldwork is not GPS dependent. Perhaps the most visible fieldwork of a surveyor is the work performed on construction sites, often before other work has begun. Unlike construction work, survey fieldwork has a very short period (if any) that is vulnerable to GPS jamming. The surveyor need only find a monument or receive one definite GPS coordinate to be able to determine everything else on site. While the non-GPS surveying may be slightly slower than pure GPS work, it is, nonetheless, a fairly rapid process on site, with most of the work stacked on the front and back ends with desk work.

For surveying, we considered the Atlanta MSA. Atlanta is in the middle of a construction boom, and surveying tends to be associated with new construction, where boundary lines of property and the location of particular utilities become very important. In Atlanta, as per the 2012 Economic Census, total receipts from surveying were $44 million annually[19] in 2012 ($49 million in 2018 dollars)—corresponding to $130,000 per day, or $190,000 per workday. There are thousands of construction permits associated with the probable use of surveying in the Atlanta MSA for the years 2017–2018. Some of these permits are associated with large building projects, which most probably feature surveying. Drawing on local construction reporting and overhead imagery, a cursory sample located dozens of major sites at the beginning of construction.

As in our analysis of construction, we used geographic analysis to identify which of the survey sites could be covered using powerful but still man-portable jammers.

However, not all of the labor associated with survey work is GPS dependent. O'Connor et al. (2019) estimated that without GPS, survey jobs would take 69.2 percent longer to do—it follows that if work hours were unchanged, surveyors would only complete 59.1 percent of their normal work during the outage, and 40.9 percent of work would not be done. Further, O'Conner et al. estimated that 30.7 percent of survey work could not be completed at all without GNSS.

We therefore estimate that jamming would prevent about 40.9 percent of survey work from being done. (Alternatively, the fraction of survey work not done could be reduced to 30.7 percent but at the cost of increased work hours.) We conclude that a single high-power portable jammer would inflict economic costs of about $18,000 per workday ($13,000 per day), while a blanket jamming campaign would inflict economic costs of about $76,000 per workday ($54,000 per day).[20]

Based on observations of online discussions, surveyors already deal with smaller solar storms in part by delaying work in the field or, if that is not possible, later rechecking their surveying in spots to see if there were errors (Frame, 2014).

Nationwide GNSS Outage

In the event of a nationwide outage due to a CME, for example, which would affect all survey work without geographical limitation, we conservatively assume that about 40.9 percent of all survey work would be prevented. From the 2012 Economic Census, annual nongeophysical surveying work in the United States was worth $6.18 billion in 2018 dollars, while annual geophysical surveying was worth $3.47 billion in 2018 dollars.

We therefore conclude that a nationwide outage could cause losses of $6.9 million per day ($9.7 million per workday) in the nongeophysical survey sector and $3.9 million per day ($5.5 million per workday) in the geophysical survey sector. If both sectors are included, the potential losses are $11 million per day ($15 million per workday).

---

[19] This value is excepting activities categorized as geophysical services, which relate to energy and mining activities rather than the construction activity that is of interest here.

[20] In other words, about 40.9 percent of the total survey activity within range of the jammers.

For comparison, the extrapolated range from Leveson (2013) and Pham (2011) (including both geophysical and nongeophysical surveying) is between $14 million and $17 million per day. O'Connor et al. (2019) arrived at a loss number of $11 million per day, excluding nongeophysical surveying which O'Connor et al. treated separately in categories for Mining and Oil and Gas.

In Table 4.1 and Table 6.1, we report both geophysical and nongeophysical surveying numbers. To avoid possible double counting, we will include only the nongeophysical surveying losses as they are to be aggregated with Mining and with Oil and Gas loss estimates from O'Conner et al. (2019). This produces an estimate of $7 million per day in 2018 dollars.

Aviation

Aviation is one of the most scrutinized means of transportation. Even relatively small airline disruptions quickly become national news, with actual accidents quickly triggering vigorous investigations and the development of procedural, hardware, and regulatory responses among the members of the aviation community. As such, the question of PNT alternatives as well as the potential effects of disruption to the PNT ecosystem are relatively well studied compared with other sectors examined in this report. Martinez et al. (2017) performed a detailed analysis on alternative PNT systems at the direction of the Federal Aviation Administration (FAA). For one piece of this analysis, MITRE sought "to examine the number of aircraft operations conducted within the vicinity of the scenario airports to identify what percentage of aircraft operations might be subject to adverse effects from a GPS outage." They complemented this analysis with another, whose goal was "to postulate a GPS outage centered on the principal airports defined in the scenarios and attempt to assess the number of aircraft adversely affected by such an outage." The MITRE team concluded that although not directly comparable, both analyses pointed to the fact that "the number of aircraft determined to be completely unable to continue navigation in the face of a GPS outage was practically insignificant based on either analysis (less than 1% for both the Nearby Airports Analysis and the IMC Flights Analysis)." These analyses point to the fact that because the aviation industry has such a robust network of alternatives to fall back on in the event of a small, localized outage, the number of affected aircraft would likely be relatively small. On the basis of these reports and the history of aviation regulation, we believe this will continue to be broadly true even while the network of alternatives shrinks as planned to lower costs. While there is some very small possibility that a GPS outage could result in loss of life in some circumstances, it is unclear that such an outage would, at present, create the potential for a widespread catastrophic scenario involving substantial loss of life.

While aircraft could thus land safely in essentially all plausible cases, there could still be disruptions of an airport if the impacted area is localized or potentially of the entire nation if GPS service were disrupted from a large-scale space event. In conversations with the FAA, it became clear that this is because currently only GPS is used to initialize IMUs, which means that commercial aircraft should not, per established procedures, leave an airport gate without GPS. However, the aircraft inertial navigation system (INS) can technically be initialized by starting from a known reference location, such as a parking spot.[21] In the future, use of the Galileo GNSS will be allowed, but no other system is currently viewed as an acceptable complement or substitute. Developing new standards for an alternative signal has proven to be a time-consuming process given the current highly conservative, safety-oriented set of processes; it has often taken roughly a decade, quite apart from the time and cost of providing any new equipment or antennas on aircraft. Consequently, in a widespread outage covering the nation, the FAA believes that workarounds would be developed to allow initialization and flight operations, albeit at some reduced rate. Without GPS, the Automatic Dependence Surveillance-Broadcast–based en route surveillance system would also be disrupted, forcing a modest increase in the spacing of aircraft to ensure safety.[22] Advance planning for such a situation might involve spreading commercial flights into the unattractive night hours, increasing capacity at the cost of convenience to the public. In any event, the air system could in an extreme case suffer costs for the duration of a nationwide outage, costs that we estimate below.

There is strong empirical evidence on which to base plausible estimates of the costs of airline disruption over a variety of different time scales. One excellent source of data for short-term disruptions comes from the recent experience(s) of Delta Airlines at its main hub of Hartsfield-Jackson International Airport (ATL) in Atlanta, Georgia. Since Hartsfield-Jackson is the busiest airport in the world, one would expect that even minor delays there could generate significant economic disruptions. In late 2017, a fire in an underground tunnel resulted in a blackout in the airport for almost 12 hours and forced planes to be grounded at the airport (Jansen, 2018a; Yamanouchi, 2017; Grinberg et al., 2017). Delta reported its pretax losses for the disruption, which took several days to fully clear, at between $50 and $60 million (Josephs, 2018; Jansen, 2018b). In April of 2017, a series of storms hit the Atlanta area, disrupting airline traffic out of ATL; the cascading series of problems, as Delta lost track of crews, reportedly cost the company a pretax total of $125 million (Breslin, 2017). Widening the scope to consider disruptions at the international level, estimates of the total losses accrued by airlines from the Icelandic volcano eruption of 2010 put the losses across the entirety of the affected area at around $1.7 billion, not adjusted for inflation (Mazzocchi, Hansstein, and Ragona, 2010; Ulfarsson and Unger, 2011). The Eyjafjallajökull Volcano was an

---

[21] Conversation with FAA personnel, December 17, 2018.

[22] Conversation with FAA personnel, December 17, 2018.

event that presented an obvious safety threat, across a very wide geographic area and persisted for weeks, longer than the likely disruptions in GPS we examined.

Abt Associates (2017) examined the threat of space weather events to aviation and suggested that the cost to airlines of actions intended to mitigate the effects of space weather could be between approximately $1 million to $30 million a day in an extreme space weather scenario. Notably, however, that prediction was based on the idea that space weather interference with high-frequency radio would interfere with air traffic communication in polar regions, causing flights in polar regions to be rerouted or canceled, and did not involve any costs from GNSS disruption. Abt Associates examined past interference events with GPS but were unable to determine if these past events had ever caused a flight to be delayed or if they were merely a nuisance forcing pilots to use alternative navigation procedures.

The estimates extrapolated from Pham (2011) suggest that the nationwide aviation industry may derive benefit of between $4 million and $9 million per day from GNSS, which would be lost in an outage. Recall that this is an extremely rough estimate, based only on the supposition that the aviation sector might get back about eight times as much benefit from GPS equipment as it spends. Nevertheless, this range seems plausible—much higher numbers would require major disruption of air traffic, and there is no evidence or expectation that GNSS denial would cause so much disruption. Order-of-magnitude lower numbers would not be very plausible either, since the aviation sector must be deriving some benefit from the GNSS equipment it buys. We therefore use the midpoint of Pham's estimate, $6 million per day in 2018 dollars, in preference to the much lower estimate in Leveson (2015a, 2015b).[23]

Maritime

Wallischeck (2016) lists 16 GPS dependencies in the maritime sector. In this section we will discuss the implications of a GNSS outage for boats and ships navigating on oceans and in waterways. The movement of goods to and from ships once in port will be discussed in a separate section.

Grant et al. (2009) tested the effects of low-powered GPS jamming against a ship underway at sea and revealed that multiple systems will report the vessel's speed and location incorrectly under such conditions, concluding that "GPS service denial has a significant impact on maritime safety" across a number of different areas. However, to be vulnerable to this type of attack, ships generally need to be relatively close to shore where both physical visual aids to navigation and a myriad of other systems designed

---

[23] Leveson (2015b) was aware that his estimate of GPS benefits in the aviation sector was probably low and that "more benefit information is needed for aviation and other transportation areas."

to provide situational awareness to the ships and to ensure safety exist. It is not clear, therefore, that attackers would have a good chance of causing widespread damage.[24]

In especially close quarters such as ports or closed waterways, such as the St. Lawrence Seaway, expert local pilots are often required, and tugs can assist ships in especially hazardous or restricted waterways (English et al., 2014). In conversations with the St. Lawrence Seaway Interest, we learned that those operating in the seaway are extremely familiar with the area.[25] Either they operate there on a regular basis as a ship's captain or muster in a licensed Seaway pilot. The Seaway is shared by Canada and the United States, and Canada provides a variety of navigation aids. In the more complicated, dangerous zone of the Thousand Islands, the Interest maintains positive control over ships, with a combination of systems. The ships themselves must have plenty of backup equipment, including multiple radar systems, steering equipment, and multiple anchors. All large commercial vessels have Automatic Identification System (AIS), a GPS integrated system, in addition to a variety of other backups. In the event of poor weather, most vehicles can continue to use radar or simply drop anchor; a wide-spanning radio broadcast goes out when visibility drops as a warning to all vessels. Smaller vessels operate without AIS or GPS, but they are familiar with the rules of the road as well as with the geography. Failure to adhere to the rules of the road is an issue of safety of life but ultimately the responsibility of the ship's captain. Given these myriad redundancies and long-established practices for safe navigation, it is unclear whether hostile actors would be able to generate significant operational and/ or economic effects without multiple, unrelated failures occurring in succession.

Moreover, in conversations with the Coast Guard, it became increasingly clear that adherence to safety standards and rules of the road were the most important element for maintaining safety of life.[26] When questioned about how the absence of GPS would impact the U.S. Coast Guard's (USCG) ability to monitor high-traffic areas, the interviewees made it clear that licensed pilots and captains are highly competent and are able to use a myriad of instruments ranging from AIS to radar, to eyes, to keep themselves and their vessels safe. When high-tech systems slow down or falter, the USCG has proven able to continue operations with little-to-no impact on traffic via a combination of physical models of movement (for situational awareness), voice traffic, and cameras. When interviewees were asked how much impact the loss of AIS could have on movement, the answer was an unequivocal "none." Bigger concerns were poor visibility in bad weather and amateurs putting themselves in harm's way.

---

[24] For another study that attempted to take control of a vessel underway in the Mediterranean, see Bhatti and Humphreys (2017).

[25] Conversation with St. Lawrence Seaway Interest, January 28, 2019.

[26] Conversation with USCG, November 19, 2018.

Recreational Boating

The study by O'Connor et al. (2019) includes both a wide range and a point estimate of damages across the maritime sector; here, we will use their point estimates. They mostly concurred that the disruption to navigation by large ships would be small, while also predicting $12 million in daily losses in commercial fishing and $2 million per day from navigation in seaways. They also predicted a much larger impact of $109 million per day in lost recreational value to boaters. This figure is a nonmarket estimate value of recreation to boaters and is not intended to correspond to an amount of money that would be spent or not spent by boaters during a GNSS outage (although, by coincidence, $109 million per day is almost exactly the total amount that is spent on all marine recreation [National Marine Manufacturers Association, 2018]).

The range of maritime estimates extrapolated from Pham (2011), which appear to also be dominated by the perceived benefit of GPS equipment to amateur boaters, are between $36 million and $85 million per day—lower than the O'Connor et al. (2019) estimate but not by a large factor. We accept the O'Connor et al. estimates above, $126 million per day in 2018 dollars, to be conservatively high in our opinion.[27]

Railway

In the railroad industry, the integration of PNT has been relatively slow and unevenly distributed. Wallischeck (2016) identified seven rail dependencies on GPS (as opposed to 11 and 16 GPS dependencies for the aviation and maritime modes), of which only five require positioning fixes. Of these dependencies, the most critical for safety of life appear to be for Positive Train Control (PTC) and potentially vehicle-to-infrastructure (V2I) communication systems. Both these systems rely on GPS position fixes to ensure safety while a train is being operated by providing awareness of the location of the train, but it is important to note that PTC is not fully deployed, and V2I communications are in development stages. The EU is exploring concepts similar to these and plans ultimately to rely on PNT systems to locate trains (Marais, Beugin, and Berbineau, 2017). In the United States, trains are currently located by "in-track transponder-based positioning," and PTC intends to mitigate the costs of additional in-track infrastructure through the use of augmented GPS (Baker, 2012). In conversation with an industry expert, it appeared that the largest threat would be designing this system to be overly reliant on PNT signals and, thus, neglecting redundant fallbacks such as those adopted in other sectors.[28] PTC fails somewhere in the system on a regular basis; opera-

---

[27] Meanwhile, the maritime benefit estimate in Leveson (2015a, 2015b) is very low, to the point where it would be hard to explain why the maritime sector spends as much on GPS equipment as it does for so little benefit. Leveson was aware that his estimate was conservative and that "more benefit information is needed for aviation and other transportation areas."

[28] Conversation with railroad professional, November 28, 2018.

tors are trained to talk with dispatchers, communicate the situation, and proceed with caution using their own judgment. From an operational perspective, this proves to be unimportant because PTC is a backup, safety-net system, used to slow or stop trains in the event of an operator error. GPS jamming would, at worst, induce a failure of PTC similar to those that could occur for other reasons; the loss of the safety net does not itself cause damage unless combined with other problems while the operator is incapacitated or failing to pay attention.

Rail, both freight and passenger, has a low fatality rate. Between 1990 and 2017, the median death rate on trains from accidents was ten per year (DoT, Federal Railroad Administration, Office of Safety Analysis, 2018). The mean is slightly higher, at 13 per year, increased by large events such as the 1993 Big Bayou Canot derailment killing 47 people (NTSB, 1994) or the 2008 Metrolink 111 crash that killed 25 people near Chatsworth (NTSB, 2010). While the Chatsworth crash was caused by operator inattention and would likely have been prevented by PTC, PTC might not have impacted the Big Bayou Canot derailment. The main dangers to a train with a functional operator are technical faults, environmental disasters, or road vehicles at railway crossings, which are not problems that PTC can solve (Union Pacific, undated).

Given the fact that rail transport does not have the same type of high-profile vulnerabilities that exist in the maritime or aviation sectors, it is not obvious how attackers would seek to generate a disruption.

For our aggregated estimate of damages, we again extrapolate the methodology of Pham (2011), which produces a very small amount, $100,000 per day in 2018 dollars.

Telecommunications

The principal wireless technology used by AT&T, Verizon, and Sprint is Long-Term Evolution–Frequency-Division Duplexing (LTE-FDD). LTE-FDD requires that the nodes in the network have good agreement on frequency, within 50 parts per billion. LTE-FDD cells may make use of GPS as a common frequency reference, although other technologies such as Precision Time Protocol (PTP) over fiber are also used. LTE-FDD does not impose a particular time synchronization requirement (Microsemi, 2014), which makes those networks much less dependent on GPS. According to information provided to DHS by the Alliance for Telecommunications Industry Solutions (ATIS) (Goode, 2018), a typical LTE-FDD macrocell has a quality oscillator and could maintain its frequency reference without GPS for a year.

LTE-TDD (Time-Division Duplexing) is a complementary technology, which allocates time slots for phones to transmit. Transmitter power being equal, LTE-TDD generally covers a smaller area than LTE-FDD but offers more flexible use of uplink and downlink bandwidth. Of the major carriers, AT&T and Verizon rely solely on LTE-FDD while Sprint uses LTE-FDD to provide baseline coverage and LTE-TDD to provide additional data capacity in high-traffic areas (Saw, 2014).[29] LTE-TDD cells have the same frequency synchronization requirement as LTE-FDD but also have a time synchronization accuracy requirement of one to five microseconds (μs), depending on the size of the cell, to avoid call interference and dropped calls between cells. In the event of a GPS outage, if no other timing reference was provided, LTE-TDD cells could drift out of sync after about a day. However, LTE-TDD makes up a relatively small part of the wireless infrastructure.

The legacy Code-Division Multiple Access (CDMA) technology, which is still used for voice calls by Verizon and Sprint, also has a time synchronization requirement of ten zs. A CDMA cell that was out of sync with its neighbors could still be used to place calls, but calls would be dropped if and when users moved across the cell boundaries (Goode, 2018).

O'Connor et al. (2019) assessed that a nationwide GNSS outage would cause significant disruptions in the wireless telecommunications sector. Their midpoint estimate was that the damage would be $40 million on the first day but that the damage per day would rise nonlinearly over time to $456 million per day after 30 days, corresponding to losing most of the value of the U.S. wireless network. The midpoint estimate of accumulated losses over a 30-day outage would be $10.4 billion.

We think that that estimate is extremely pessimistic, however. First of all, it is not clear that the majority of the U.S. wireless network would even be affected by GPS outage of even a few weeks. For example, Goode (2018) states that the dominant cellular technology in the United States, LTE-FDD, which is used by AT&T, Verizon, and Sprint, can support normal voice and data services for weeks or months without GPS. Second, even if the network were affected, the baseline ability to make wireless calls would not be threatened. Individual users would have immediate, simple workarounds for degraded service: if a call was dropped, hit redial; if a call is repeatedly dropped, pull over and complete the call while parked; if interference is degrading a high-data-rate connection over LTE-TDD, switch to a Wi-Fi connection. Third, the U.S. wireless carriers would have strong incentive to work to restore any degraded service, if it occurred, and would have many options for resynchronizing their networks in order to do so, whether over fiber, by RF signal, or by physically transporting atomic clocks. For these reasons, we assess that damages to the telecommunications sector would likely be very small and do not consider them further in our aggregation. This is also supported by the historical evidence discussed at the end of this chapter.

---

[29] As of late 2016, LTE-TDD service was commercially offered in the United States by Sprint, by DISH Network in Corpus Christi, by SpeedConnect in the Quad Cities, and by Redzone Wireless in Maine (GSA, 2016).

Electricity Generation and Transmission

Although we did not perform a detailed analysis of the electrical power sector, our initial investigation indicated that the use of GPS in phase measurement units, for synchronizing the electric grid and for locating physical failures through precise timing, would not be significantly affected either by local jamming or by a general GPS disruption. Spoofing, which might produce other effects, can be essentially eliminated by taking advantage of existing alternative signals, including a local clock and the quite fixed location of the phase measurement unit.

O'Connor et al. (2019) examined the electrical sector in more depth. They report that

> in the event of a 30-day failure of GPS today, a major or prolonged failure of the electrical system is unlikely because of safeguards and backup contingency plans in place. However, the probability of outages might increase. In addition, faults occurring from natural or non-natural events would take longer to identify and repair, increasing the duration of outages. . . .

> PMUs [phase measurement units] themselves have a built-in backup feature that enables them to maintain holdover for several days. Although this holdover will start gradually degrading after several days, it will give utilities enough time to adjust to the absence of a standard global time by, for example, manually calibrating the time. (O'Connor et al., 2019)

They estimated that the increased probability and duration of outages would result in a point estimate of the costs of $9 million per day, which is also the amount in 2018 dollars. We adopt this estimate.

Finance

The finance sector is one area that has taken advantage of GPS for absolute timing.

Many financial transactions rely on PTP or Network Time Protocol (NTP). Those reliant on PTP often use GPS as a form of confirmation of timing and clock synchronization. Rapid trades depend on knowing who held which stocks at any given time and on properly sequencing transactions. When trades occur in the course of milliseconds, this knowledge depends heavily on precision timing. GPS currently serves as a simple way to get that precision timing. Although the sector has adopted GPS as a source of timing information, analyses done within the sector have acknowledged both the potential for GPS to be subject to disturbances at both the receiver location and from satellite services themselves and thus the need to maintain backup sources of timing to protect the capability to execute timing-sensitive transactions (that also require precise and widely coordinated time stamps) through a period of such disruption. The financial industry uses a variety of systems such as atomic clocks as backups in case GPS signals are disrupted; such local backups would also allow a firm to operate through most realistic jamming incidents that might be expected to measure in at most days.[30] For example, as of 2017, the operator of the New York Stock Exchange was working to use the U.K. National Physical Laboratory's atomic clock for precise timing at its international datacenter in Basildon, United Kingdom ("NPL Offers Precise Timing for Intercontinental Exchange Data Centre in Basildon, UK," 2017). Such an approach to using off-site timing is indicative of the financial industry's commitment to have multiple options for time services.

These backup features and attendant fallback infrastructure require that financial industry firms spend some money, but given the concentration of resources in the financial industry and the time sensitivity of revenue flows, market forces create an incentive for each individual firm to have backups in place to maintain operations through GPS disruptions. Therefore, given the potential for maintaining local backup timing, estimating the potential costs of jamming GPS timing signals would first involve the probability of a given firm having put a backup in place, the probability of that backup failing when called on, and the amount of money processed by the firm on any given day, averaged out across the year.

Importantly though, the financial effect is *disruption*, not a direct loss; if a transaction is not executed, there is some opportunity cost to the seller and buyer and a lost fee to the entity closing the sale, but the seller still has the asset and the buyer still has its funds. Additionally, if there were any localized disruption, the most probable effect would be simply to shift to another exchange or bank that was not disrupted, leading to even less loss. In fact, it is precisely that shifting that provides the private entities their incentive for internal timing backups.

Regional Jamming Vignette: Los Angeles

For a geographically constrained case that would reflect the practical challenge of an attacker covering multiple financial firms with GPS jamming, we selected Los Angeles as a city with a significant and concentrated financial sector. In the overall financial services sector, Los Angeles is third after New York and Chicago. We took the sector GDP of finance located solely in Los Angeles and then looked at a sample of major financial institutions in the region, the majority of which are geographically clustered in the downtown area.

According to BEA (2018b), the sector GDP of finance, located in Los Angeles and associated with securities (and therefore potentially with timing requirements for reconciling trading activities) was $10.1 billion in 2016 ($10.5 billion in 2018 dollars).

---

[30] It should be noted that in timing applications, relatively brief windows of reception are necessary to obtain the shared time and allow for coordination of local clocks. See Lombardi et al. (2016) for a discussion of the use of clocks in the financial industry, holdover performance, and synchronization requirements/approaches.

Figure 4.4
View from Scenario Jamming Location Showing Relevant Downtown Built Environment in Jammer Field of View

SOURCE: Google Earth Pro, version 7.3, 2019.

According to the 2012 Economic Census, the total annual revenues associated with finance involving securities was $515 billion in 2012, while according to the BEA, the total U.S. GDP in the sector in that year was $233 billion. If we assume that the same ratio of revenues to GDP holds, we would infer that the total revenues associated with finance involving securities in Los Angeles are about $23 billion dollars annually or about $90 million per workday.

Moreover, most of the value of financial activity is concentrated in a small number of firms, and a significant number of financial firms in Los Angeles are located in a small number of buildings.

A jamming attack would not necessarily affect all GNSS receivers in the area, because some might be protected from the jamming by intervening buildings or other terrain. Nevertheless, suppose that a jammer successfully affected 90 percent of the GNSS receivers in the financial district. We conservatively assume that firms on average have backups that are operational 99 percent of the time.[31] This would produce an expected value for daily transactions disrupted by jamming in Los Angeles of approximately $900,000 per day.[32] Of course, the real cost to society is unclear since, as a result of this localized attack, transactions would shift to banks outside of Los Angeles.

---

[31] This is not a particularly good level of availability and would be expected of any managed computer system (Gray and Siewiorek, 1991).

[32] Twenty-five billion dollars in sector revenue in Los Angeles over a 260-workday period, multiplied by 90 percent jammer coverage, and assuming a one in 100 probability of backup timing failure.

Nationwide GNSS Outage

We next consider the case where, due to a CME, the entire nation's entire financial industry is denied GNSS timing. On the other hand, instead of assuming all firms have only mediocre timing backups, we assume that both the probability that a firm maintained a backup and the probability of that backup failing would be related to firm revenue, with the wealthiest firms having the strongest level of preparedness and the highest quality backups while progressively smaller firms would be less and less likely to invest in reliable timing backups. We assumed that the largest firms would all have very reliable backups, with only a one-in-a-million probability of being unavailable at any given time, and that for firms outside the top 50, the chance of backup failure trends down to a conventional system which has a 1-percent probability of being unavailable at any given time.[33]

Data are available on the size of financial firms by revenue, with the top 50 financial firms accounting for more than half of revenues. The concentration of the revenue of the top four firms was approximately 17 percent of the total value of the sector or about $370 million in revenues each workday (assuming 260 workdays per year). The top five to eight firms added an additional $200 million in revenue per workday, while firms 9–20 added another $340 million. Finally, the top 21–50 firms added $320 million in revenue per workday. All of the remaining firms combined provide $930 million per workday (U.S. Census Bureau, 2012—figures adjusted to 2018 dollars).

Given the distribution of activity among firms and our estimated backup reliabilities, the amount of money the subsector-as-a whole would expect to have disrupted from an attack of this type can then be modeled as an expected value, contingent on an attack or incident occurring. Given the daily revenue shares and assumed probability of backup failure for different firm classes—*even if an incident occurs where the entire financial industry in the country is affected by GPS jamming or disruption*—the expected disrupted transaction revenue is approximately $9 million per day, the majority of which is concentrated in firms outside the top 50 by revenue.

Again, the $9 million per day figure does not represent an expected loss to society as a whole, since any transactions prevented at a bank or exchange where the timing backups had failed would likely be rerouted to another entity, with little loss to society as a whole.

O'Connor et al. (2019) considered the financial sector and reached a similar conclusion, finding that "sector representatives do not view an observable loss of GPS for 30 days as having a substantial economic impact." We will use our small estimate of $9 million per day for aggregating damages.

---

[33] In the terminology of Gray and Siewiorek (1991), these extremes of backup performance correspond to a "very-high availability-system" and a conventional "managed system," respectively.

Port Operations

Given the relatively short range of GPS jamming systems, compared with the vast size of the open ocean, an adversary's best chance to cause significant effects would be to jam or spoof vessels attempting to enter ports. At worst, this could generate a complete shutdown of the port; estimates of such potential damage exist (e.g., Park, 2008; Park et al., 2008).[34]

The ports that might be affected are those that handle containerized cargo, as GPS systems are used to track the off-loading and storage of containers; in contrast, bulk and break-bulk shipping, as well as roll-on/roll-off vehicle cargo would be little affected. For containerized cargo at ports with systems that rely on GPS, the loss of GPS would typically slow handling of the cargo significantly.

Park (2008) considered the effects of a one-month closure of the Los Angeles/Long Beach port because of a dirty bomb attack; the study suggested that the total losses for this shutdown across the entire nation would be approximately $40.4 billion for the month in 2018 dollars, or on the order of $1.35 billion a day. However, one particularly telling critique of studies of this type are that they typically do not account for any efforts at resiliency or mitigation. A separate study by Rose and Wei (2013), examining the effects of a hypothetical 90-day port shutdown to Port Arthur/Belmont, suggested regional losses could be as high as $14.1 billion in 2018 dollars (or roughly $157 million a day), but when resiliency measures were taken into account, those numbers could shrink by almost 70 percent.

While at first glance these numbers may still seem to be worryingly high, it is important to keep in mind both the *scale* of the analysis, the *duration* of the analysis, and the *impacts of resiliency*. The figures presented by Park et al. (2008) seem enormous, but represent closing a port. If the effects of resiliency proposed by Rose and Wei (2013) are factored in, this number drops to $351 million a day in 2018 dollars, spread out over the entire country, from the closure of these two adjacent ports. Since Los Angeles/Long Beach handles about 40 percent of the total U.S. container volume (Cushman & Wakefield, 2019, O'Connor et al., 2019), this would imply a daily loss of $880 million with the resiliency correction.

Finally, Hall (2004) casts even more doubt on economic impact studies of this type, suggesting that estimates of this variety both treat the entirety of the cargo as simply lost, as opposed to merely delayed, and ignore short-term substitution behavior, leading impact studies to grossly overestimate losses. Indeed, many of these ports can expand their hours of operation, particularly over short periods, to compensate for decreased efficiency. Consequently, when thinking about short-term disruptions on the order of days, such as those considered by this report, *we believe that short-term sub-*

---

[34] However, port shutdowns can often be anticipated while terrorist attacks cannot. This is important because events that can be anticipated present opportunities for mitigation, such as diverting cargo traffic to nearby ports, or front-loading orders.

*stitution and resilience effects would likely mitigate an overwhelming majority of the eco-
nomic losses associated with any potential losses of efficiency at the port.* Indeed, as long as
substitutes are available and supply chains have factored in the possibility of delays on
shipping from a host of other issues, downstream effects are likely to be quite minimal
in the short term. This makes an estimate that is highly uncertain, between $0–$880
million per day, though as just noted, we believe the actual amount would be low.

O'Connor et al. (2019) estimated disruption costs at ports: their point estimate
was $6.7 billion for a 30-day outage or an average of $224 million per day over that
30-day period (but varying nonlinearly with the duration of the outage, with losses
escalating due to a growing backlog of containers at the ports).

There are some reasons to think that this estimate is pessimistic, however. First,
while GPS tracking and positioning of containers presumably eases employee work-
load, generally speaking, semiautomated cranes with GPS-assisted guidance options
can revert to fully manual operation. Container tracking software, such as a Navis
terminal operating system, permits container locations to be entered manually. While
employee efficiency would be reduced, most large ports do not operate a full two
shifts a day, let alone three. Therefore, by extending hours of existing workers during
a short GNSS outage and bringing on new workers during a long GNSS outage, ports
could avoid devastating port congestion and backlogs of containers at the expense
of increased labor costs. Second, supply chains have adapted since the ten-day-long
2002 shutdown of West Coast ports, precisely to avoid the effects of such a disruption
(Stevens and Ziobro, 2015). Both these factors would significantly reduce the damages
suffered for any period of disruption.

Nevertheless, we will use this estimate of O'Connor et al. in our aggregated
estimate of damages. One important note from that estimate—fully 81 percent of
the damages estimated by O'Connor et al. (2019) come from only four ports—Long
Beach, Los Angeles, New York, and Norfolk. The estimate from Cushman & Wakefield (2019) confirms a great concentration in total container volume, with 61 percent
in the top four container ports and 75 percent from the top six. Whether four or six,
this relatively small number of ports where any effect would be concentrated has direct
relevance when considering various systems that might mitigate these losses.

Mining

O'Connor et al. (2019) provided the only benefit analysis we have of the mining sector,
where GPS is used to enhance efficiency in open pit mining, using accuracies for some
functions of order one centimeter (cm) and others of three to five meters. They estimated the losses due to reduced mining productivity during a nationwide GNSS denial
to be about $32 million per day, in 2018 dollars, which we will use in our aggregation.
We infer from the reported comments of their interviewees that the majority of the

productivity difference is attributable to high-precision uses of GPS (i.e., GPS real-time kinematic [RTK]), and a PNT alternative that was substantially less accurate than GPS RTK would only restore a minor fraction of the lost productivity.

Oil and Gas Exploration

O'Connor et al. (2019) analyzed the oil and gas exploration sector and found the most critical dependence, which accounted for most of the estimated losses, to come from disrupting the production of wells in deep water where they cannot be securely anchored to the ocean floor. Differential GPS is used to produce the needed decimeter accuracies. Their estimated loss, from production that does not occur while the float-ing rigs are disconnected for their 30-day outage, was $1.5 billion.

While these losses may be correct, we note that they will have no effect on the question of backups or complementary systems, as only an augmented GNSS can offer such accuracies over the ocean; other wide area systems cannot. Thus, whatever the losses, they do not affect our analysis further.

Multisector Regional Jamming Vignette: Los Angeles–Long Beach

We now combine the above results into an example regional scenario where all sectors are threatened at once. Consider a major jamming attack on the Los Angeles–Long Beach–Anaheim MSA which covers 25 percent of the MSA area.

The second column in Table 4.4 indicates the probable nationwide losses at risk in each sector. The third column indicates three-digit NAICS codes, if any, that most closely correspond to each critical sector. The fourth column contains the share of the national sector activity that occurs in the Los Angeles–Long Beach–Anaheim MSA. This is estimated by taking total sector receipts in the MSA as a fraction of national sector receipts, according to the 2012 Economic Census (U.S. Census Bureau, 2016). If no three-digit NAICS code applies to the sector or that data is unavailable, then the MSA share of all receipts across sectors is used. Multiplying the second and fourth col-umns gives the total potential costs at risk in the Los Angeles MSA. The sixth column gives the assumed level of coverage of each sector—we could assume that different sec-tors were targeted to greater or lesser degrees, but for this example we will suppose that all sectors are 25 percent covered by jammer footprints at any given time. (It might be that the attackers are moving their jammers randomly about the MSA in order to delay capture.) Finally, the seventh column shows the estimated losses per day. In this example, the projected losses from the scenario are only around $11 million per day, with over half coming from the presumed disruption in the ports of Los Angeles and

Table 4.4
Estimates of Daily Costs from an All-Sector Regional Jamming Attack (millions of dollars)

Sector	Estimated National Sector Cost Per Day ($Million)	Related NAICS 3-Digit Codes	Estimated Share of Sector in Los Angeles–Long Beach–Anaheim MSA	Potential Regional Costs Per Day ($Million)	Assumed Jammer Coverage of Sector	Expected Losses Per Day ($Million)
Consumer location-based services	95	—	4.7%	4.5	25%	1.1
Commercial road transport	138	484 485	2.7%	3.7	25%	0.9
Emergency services	11–72	621	4.7%	0.5–3.4	25%	0.1–0.8
Agriculture	a	11	0.3%	0.2	25%	0.0
Construction	24	23	3.3%	0.8	25%	0.2
Surveying	7	23	3.3%	0.2	25%	0.1
Aviation	6	481	6.8%	0.4	25%	0.1
Maritime	126	713	6.3%	7.9	25%	2
Railway	0.1	488	9.8%	0.0	25%	0.0
Telecom	0	517	5.3%	0.0	25%	0.0
Electric grid	9	—	4.7%	0.4	25%	0.1
Finance	9	523	5.3%	0.5	25%	0.1
People tracking	0.7	—	4.7%	0.0	25%	0.0
Port operations	229	488	9.8%	22	25%	5.6
Mining	32	212	0.1%	0.0	25%	0.0
Oil and gas	a	211	0.7%	0.4	25%	0.1
TOTAL						**10–11**

SOURCES: Authors' analysis and U.S. Census Bureau, "2012 SUSB Annual Datasets by Establishment Industry," Census.gov, October 3, 2016.

a The agricultural losses analyzed come from large area crops, such as corn, and the oil and gas losses from drilling at sea, neither of which are found in this area. We thus set the cost per day here to zero, even though other parts of those sectors are present, as this table indicates.

Long Beach and the rest concentrated in disruptions in recreational boating and in personal and vehicle navigation devices.

Some Historical Evidence Against Predictions of Extreme Disruption

Significant disruptions to major transportation networks that have occurred (e.g., the complete shutdown of air transport after September 11 terrorist attacks) have generated significant economic costs when whole systems have been affected simultaneously. However, smaller-scale disruptions, which are experienced by transportation networks and other systems frequently as a result of weather and other events, have smaller-scale effects. The experience with routine disruptions suggests that it is unlikely that small-scale, short-term disruptions of GPS could trigger cascading failures in the system unless combined with other larger-scale kinetic attacks.

Large-scale disruptions are nonetheless sometimes posited (Glass, 2016; O'Connor et al., 2019; Sadlier et al., 2017). This section provides some historical context from existing disruptions, which offer a natural experiment to reveal any widespread or significant effects. We find the evidence from these actual GNSS denial events to be consistent with a denial of GNSS benefits, but inconsistent with extreme systemic disruption.

San Diego

In January 2007, the U.S. Navy accidentally jammed GPS in downtown San Diego for two hours (DHS, 2011). The incident is often mentioned as an example of the GPS jamming threat to telecommunications networks (e.g., ATIS, 2017). Curry (2010) wrote, "The well known 'San Diego incident' US DoD own-goal illustrates how localised GPS jamming can bring down a mobile telecom network."

While the jamming did deny a GPS signal to 150 cell towers and did affect a hospital's mobile paging system, the overall impact of the incident on the mobile network seems to have been exaggerated. It received no media attention in San Diego at the time it occurred. Four years later, after the U.K. publication *New Scientist* wrote a dramatic article about the incident called "GPS Chaos" (Hambling, 2011), Voice of San Diego investigated and determined that the story was misleading. Cell phone network operators lost GPS signals from downtown cell towers but switched to other synchronization methods and did not report any loss of cell service to customers (Dotinga, 2011).

Finnmark

In late October and November of 2018, commercial aircraft for the SAS and Widerøe airlines experienced loss of satellite navigation signals in northeastern Norwegian airspace, presumably because of Russian military jamming; the jamming coincided with the large NATO exercise Trident Juncture taking place in Norway. The Norwegian Civil Aviation Authority noted the disturbances, but Widerøe stated, "There are no security risks, we have good routines and this is not the first time we have experienced loss of signals." GPS jamming had also been observed a year earlier in connection with Russian military exercises (Nilsen, 2018). In these jamming situations, the Norwegian airlines had to rely on other navigational methods, but as far as was reported, there was no significant disruption of air traffic.

Incheon

North Korea has engaged in multiday GPS jamming campaigns against South Korea on multiple occasions: for four days in August 2010, for 11 days in March 2011, for 16 days beginning on April 28, 2012, and most recently for six days beginning on March 31, 2016. During the 2016 campaign, 1,007 airplanes, 715 ships, and 1,786 cell phone base stations reported being jammed. The jamming impact was felt in the city of Incheon and in Gyeonggi and Gangwon provinces (Yonhap, 2016a, 2016b).

Some of the hundreds of fishing vessels that were jammed have had to return to port.[35] Air flights have continued on schedule using other means of navigation, and there have been no accidents or disruption of operations, though at least during the 2012 campaign, there were a few incidents of pilots receiving distracting signals such as false ground proximity warnings ("N. Korean GPS Jamming Threatens Passenger Planes," 2012).

Korean wireless telecom providers have been sufficiently concerned with the GPS jamming that they have taken some reactive steps, such as trying to move or shield GPS antennas ("Seoul's Makeshift Answer to N. Korean Jamming Attacks," 2011). However, there have been no major disruptions of telecommunication systems (Kim, 2016).

By a simple analogy with the prediction by O'Connor et al. (2019) for U.S. wireless networks, one might expect that by day six of the 2016 jamming campaign, the wireless network in Incheon would have lost 40 percent of its functionality—but this did not take place.

By simple analogy with the prediction by O'Connor et al. (2019) for U.S. ports, one might expect that the Port of Incheon, the second largest sea port in Korea, would have been largely shut down for the first five days in April 2016. Instead, in April 2016, the Port of Incheon handled 223,126 twenty-foot equivalent unit (TEU) of container cargo, setting a monthly port record and exceeding the previous April by 11 percent[36] (Incheon Port Authority, 2016).

There are also no reports of massive traffic jams in Incheon, as Sadlier et al. (2017) predicted would occur during a five-day GNSS outage in the United Kingdom.

We can infer the actual impact of the GPS outage on the Korean people—moderate, but not extreme disruption—from the fact that the Korean government has invested

---

[35] However, there was no noticeable impact on the overall Korean fishing industry. Fishery income of Korean fishery households rose 15.2 percent in 2016 (Statistics Korea, 2017). The South Korean fishing fleet has over 60,000 boats, mostly operating from overseas fishing bases in all oceans (FAO, 2017).

[36] By container volume, this makes Incheon busier than all but three U.S. ports (Los Angeles, Long Beach, and New York).

some spending, but so far only $14 million, in seeking PNT backups, although it is entirely plausible that the North Korean jamming could resume at any time (Kim and Saul, 2016). As of early 2018, some sites for the eLoran deployment had been identified (Son et al., 2018). Apart from any strictly economic impacts of lost PNT, South Korea has an interest in ensuring that boats do not accidentally enter North Korean waters.

It is fair to observe that the Korean jamming attacks and other actual GNSS denial attacks have only been regional, whereas O'Connor et al. (2019) and Sadlier et al. (2017) were asked to consider a national or global outage of GNSS. On the other hand, this shows that even a rogue nuclear-armed nation-state acting with near impunity cannot easily inflict nationwide GNSS denial, even on an immediately neighboring state.

Summary

Approaching the assessment of the potential losses from GPS disruption from a nationwide perspective, our estimates—based on existing literature and parametric approaches for considering existing complements and backups—range, on a daily basis, from a low estimate of $785 million to a high value of $1,318 million. While such numbers would be consequential, there are two important factors that significantly affect how this number should be understood.

First, as the next two chapters show in detail, the alternative systems available today in general cannot prevent all these damages. In considering any investment in an alternative system, only those damages it would actually prevent are relevant.

Second, most GPS disruption scenarios—including jamming and other disruption by adversaries across the threat spectrum that we discussed in Chapter Three—would not affect the entire country. Focusing on daily costs for a single metropolitan area and on the industry sectors there that use PNT services results in estimates in the millions of dollars per day.

Because nationwide analysis is by definition done from the top down, it necessarily neglects some specific details like geography or specific responses to GPS disruption by affected individuals or organizations that are nonetheless critical for assessing the consequences of such incidents. As a result, we complemented those calculations with analyses from the bottom up, exploring how such factors could affect the consequences of GPS jamming incidents. Such analyses also better reflect the scale of disruption that malicious actors can realistically hope to inflict in pursuit of whatever goals they believe that PNT disruption might advance. Prior attempts at costing GPS disruptions have rarely taken such a bottom-up approach, and so our attempt represents an effort to establish better likely order-of-magnitude costs for disruptions of this kind. As can be seen throughout this section, the estimates of damage can sometimes vary considerably. Despite this, we believe that this bottom-up approach is more likely to yield estimates that more closely approximate reality, certainly in small-scale scenarios and arguably when aggregated nationally, than the maximalist assumptions previously adopted in analyses of this kind.

At the highest level, these regional analyses show that the direct costs of small-scale, short-term disruptions are likely exactly that—small scale and short term. This is driven by the fact that the range of GPS jammers is relatively small compared with the size of both the country and the economy. When combined with the fact that the persistence of effects is also short (i.e., no physical damage is actually done), it is difficult to see how GPS jamming alone could produce long-lasting effects that compound over time. Even in a case where multiple sectors are concentrated, such as in major cities or around transportation hubs, jammers that may be well located for one type of attack may not be well located for another type of attack.

Another major takeaway of our analyses is that other components of the PNT ecosystem—including specific mitigation measures to the loss of GPS—can play a role in reducing the consequences of disruptions to GPS. While GPS has contributed large amounts to the economy, those who has ever had the battery die on their cell phone while attempting to drive to a destination have personally experienced GPS being unavailable and falling back to alternative methods of positioning and navigation. As previous economic analyses focusing on the economic costs of events like natural disasters and transport disruptions have demonstrated, mitigation strategies may be as simple as rescheduling work or accepting reduced efficiency. In other cases, prudent economic actors build buffers into their operations because delays occur for any number of reasons, most of which have nothing to do with the availability or unavailability of PNT services. Where the economic or safety consequences are extremely high, such as in aircraft landings, narrow waterways, or the financial services sectors, multiple backups are already in place. Perhaps the real danger would be in a future overreliance on a single system/backup or the elimination of safety margins in pursuit of efficiency improvements or marginal cost reductions.

CHAPTER FIVE

Identify Alternative Sources and Technologies to Increase National PNT Resilience and Robustness

Because of the importance of PNT in the modern economy, a wide range of technologies have been proposed that could supplement or back up GPS as a keystone component of the system. At one end of the technology spectrum are systems intended to cover wide areas or the entire nation. Enhanced Long-Range Navigation (eLoran) is one such system that has been proposed as an addition to explicitly back up GPS in response to the threats described in Chapter Three. A variety of other technologies could also supplement the PNT ecosystem in different ways by providing alternative capabilities for PNT or by performing other functions; such other functions include integrity monitoring of GPS signals to detect disruption (and therefore defeat attempts as spoofing by warning users it was occurring) or sensors to detect the transmission of jamming signals to rapidly halt such incidents and enable the perpetrators to be apprehended. Many of these alternative PNT technologies or supplemental systems already exist and are thus readily at hand but are not widely implemented. Others are at earlier stages of development.

Because different critical infrastructure sectors and industries have different applications for PNT data, the level of accuracy and precision that is useful to them differs. As a result, some of the potential alternatives or complements do not meet all industries needs and thus may be more or less attractive as additions to the national PNT ecosystem.

This chapter addresses the third part of the analysis: ***identify alternative sources and technologies to increase national PNT resilience and robustness.*** In doing so, it answers the following questions: What are the mature technologies/systems that could supply alternative PNT signals? How is each characterized—by area covered, service provided (timing only or full PNT), accuracy, applicability to different users (cell phone compatibility, requirement for fiber connection, etc.), vulnerability to various threats (importantly those sharing vulnerabilities with GPS, such as other GNSS), and cost? ***Our bottom line answer to those questions is that there are many alternatives but no single perfect system. The other GNSS constellations could provide seamless backup for GPS but are vulnerable to the same major events (i.e., space storms) that could disrupt GPS. Proposed terrestrial backups to GNSS constellations can either provide wide area coverage or high-accuracy positioning but*** *not both. As such, a diverse set of PNT sources is better and more robust than any single system.*

In the remainder of this chapter, we briefly discuss the approach we took to identifying alternatives sources and technologies and then explore those sources and technologies.

To explore this space of potential levels of PNT capability across different industries, we first reviewed data collected directly that explore stated user needs of accuracy and precision across industries. Cavitt et al. (2018) surveyed users for their precision timing requirements while Tralli et al. (2018a, 2018b) identified user needs for positioning and navigation.

We perceive that these user needs can be divided into five categories.

A. **Precision Timing on Networks.** Modern communication networks, broadly defined to include financial networks and electrical power networks, desire different nodes to have a common time standard so that network operations can be tightly synchronized. GPS is widely used today. Indoor reception of a timing signal would be valuable to avoid the need for a rooftop antenna. Telecom operators are moderately sensitive to the cost of extra equipment (ATIS, 2017).

B. **Moderate-Accuracy Positioning of People and Road Vehicles.** These user needs are fulfilled today primarily by smartphones and similar devices. Realistically, mass adoption of a new system for these users will only occur if it can be integrated into smartphones at low cost. By analogy with cell phone service, it might be acceptable to cover 99 percent of the population without providing truly global service.

C. **High-Accuracy Positioning of Equipment.** These users need submeter or even centimeter-level precision for surveying and the guidance of expensive equipment for construction, mining, drilling, or agriculture. Today they use differential GPS and are accustomed to paying for high-end PNT services. In many cases, these users are located in remote areas.

D. **High-Reliability Positioning in Aircraft Landing.** Aviation users have precision requirements that are demanding in a slightly different way. Submeter accuracy is not required, but the error must be bounded with very high reliability; an occasional error of many meters would be unacceptable. For landing aircraft, vertical position accuracy is important, which is difficult for some PNT systems to deliver.

E. **Low-Accuracy Positioning of Aircraft and Ships En Route.** Finally, "classic" navigation users do not need indoor reception, can afford user equipment beyond a smartphone, and do not have highly demanding accuracy requirements—but they do need coverage across wide areas, possibly far from shore.

The reported accuracy needs of users in these five areas are shown in in Tables 5.1 through 5.5.

Table 5.1
Reported User Needs for Timing Accuracy on Networks

Critical Sector	Application	Accuracy Range
Communications/ mobile applications[a]	Billing, alarms	1–500 ms
	Internet protocol delay monitoring	5–100 µs
	Call handoff/continuation[a]	10–30 µs
	Node-to-node communication[a]	7–9 µs
	Network routers and switches, network backhaul[b]	4–5 µs
	Time stamping/event management	2–5 µs
	LTE-TDD (large cell); WiMAX-TDD (some configurations)	1.5–5 µs
	UTRA-TDD; LTE-TDD (small cell)	1–1.5 µs
	Handoffs in WiMAX-TDD (some configurations)	1 µs
Wired communications[c]	Conversational video (live streaming)	150 ms
	Conversational voice	100 ms
	Mission-critical data	75 ms
	Mission-critical delay-sensitive signaling	50 ms
	Vehicle-to-everything messages	50 ms
	Network routers and switches[b]	1.5 µs to 50 ms
	Grandmaster clock[b]	1.5 µs to 50 ms
Electricity	Physical/video security	1 s
	Network security	1 ms
	Sequence-of-event recorder	1 ms
	Protective relays—coordinated controls	1 µs
	PMU—offline	1 µs
Emergency services	Public safety answering point	Sub–1 s
	Simulcast Land Mobile Radio systems	2 ms
	FirstNet	1.5 µs
Financial services	Manual security trading	1 s
	Automated security trading	50 ms
	Computer system clocks and time stamping	100 µs to 50 ms
	Non–high-frequency trading (ESMA MiFID II)	1 ms
	High–frequency trading (FINRA CAT NMS, ESMA MiFID II)	100 µs

SOURCES: Cavitt et al., 2018; FINRA CAT, 2020; European Union, 2016.

NOTES: ESMA = European Securities and Markets Authority; UTRA = Universal Mobile Telecommunications System Terrestrial Radio Access; PMU = phasor measurement unit. FirstNet = First Responder Network Authority. FINRA = Financial Industry Regulatory Authority," "CAT = Compliance Audit Trail," "NMS = National Market System

[a] Applies to second generation (2G), third generation (3G), LTE frequency-division duplexing, and LTE Advanced, LTE-FDD, and LTE-A, except where otherwise noted.

[b] Violation of a timing requirement expected to have a relatively minor impact.

[c] Applies to Synchronous Optical Networking (SONET), time-division multiplexing (TDM), Ethernet, and ultra-high-speed Ethernet.

Table 5.2
Reported User Needs for Moderate-Accuracy Positioning of People and Road Vehicles

Critical Sector	Application	Position Accuracy Range
V2I applications	Road	5 m
	Lane	1.1 m
	Where in lane	0.7 m
V2V applications	Road	5 m
	Lane	1.5 m
	Where in lane	1.0 m
Chemicals	Tracking chemicals through a supply chain	1–5 m
	Inspection and monitoring of equipment, pipes, and assets	Sub–1 m
	Chemical cleanup	Sub–1 m
Commercial facilities	Location-based marketing and sales	Sub–5 m
	Geographical service extension	Sub–5 m
Communications	Wireless signal strength measurement	Sub–5 m
	Service and fleet management	Sub–5 m
	Public safety alert management	Sub–10 m
Emergency services	Strategic deployment of resources (large incidents)	Sub–1 m
	Dispatch and routing (routine incidents)	Sub–1 m
	Public safety alert management	Sub–10 m
Financial services	Tracking assets, such as cash	15 m
	Risk assessment	5 m to track consumer auto behavior
	Loan loss mitigation/measurement	5 m to track automotive collateral
Food and agriculture	Food sourcing	5 m
	Food control	Sub–5 m
	Workforce/asset tracking	Sub–5 m
Government facilities	Base planning/coordination	Sub–1 m
	Student tracking systems	Sub–5 m
	Defendant/parolee tracker	Sub–5 m
Health care and public health	Health data mapping	Sub–1 m
	Location-based services to direct patients to health services	Sub–5 m
	Telemedicine and response	Sub–5 m
Nuclear materials and waste	Tracking materials and waste through a supply chain	1–5 m
Water and wastewater	Fleet management	Sub–5 m

SOURCES: Thompson, Chen, and Lawson, 2018; Tralli et al., *Annex: Transportation Sector, Positioning and Navigation Critical Infrastructure Market Assessment and User Needs Framework and Methodology: Towards Developing a Formal Set of Structured, Validated Requirements and Technical Specifications for a Complementary or Backup System to the U.S. Global Positioning System*, El Segundo, Calif.: Aerospace Corporation, November 30, 2018b.

NOTES: V2I = road vehicle to infrastructure; V2V = road to vehicle.

Table 5.3
Reported User Needs for High-Accuracy Machine Positioning

Critical Sector	Application	Position Accuracy Range
Commercial facilities	Construction	Sub–1 m
Dams	Monitoring deformations in dams and infrastructure (structural integrity)	1 cm horizontal 2 cm vertical
	Monitoring deformations in landforms and waterways	Sub–0.5 m
	Construction of dams	Sub–10 cm
Energy	Seismic exploration (land and marine)	1 m for seismic exploration 10 cm for hydrographic mapping
	Dynamic positioning—drilling at sea	1 m for docking
	Construction	10 cm for dredging and construction 1 m for cable and pipe laying
Food and agriculture	Mapping farms	Sub–1 m
	Piloting farm equipment	Sub–1 m
	Variable-rate technology	Sub–1 m
Maritime	Port	1 m
Nuclear reactors, materials, and waste	Inspection and monitoring of facilities	1 cm
	Monitoring crustal deformations at a nuclear waste disposal site	Sub–1 m
Water and wastewater	Equipment mapping, monitoring, and tracking	Sub–1 m
	Survey and mapping of landforms and waterways	1 cm

SOURCES: Thompson et al., 2018; Tralli et al., 2018b.

Table 5.4
Reported User Needs for High-Reliability Positioning in Aircraft Landing

Critical Sector	Application	Position Accuracy Range
Aviation	APV	16 m horizontal 20 m vertical (APV-I)
		8 m vertical (APV-II)
	CAT I landing	16 m horizontal, 4–6 m vertical
	CAT II landing	7.5 m horizontal, 1 m vertical
	CAT III landing	3 m horizontal, 1 m vertical

SOURCE: Tralli et al., 2018b.

NOTE: APV = Approach Procedure with Vertical Guidance.

Table 5.5
Reported User Needs for Low-Accuracy Positioning of Aircraft and Ships En Route

Critical Sector	Application	Position Accuracy Range
Aviation	Oceanic phase of flight	7.4 km
	En route flight	3.7 km
	Terminal flight	750 m
	Non-precision approach (NPA)	220 m
Maritime	Ocean navigation	10 m
	Port approach and restricted waters	10 m
	Inland waterways	2–10 m

SOURCE: Tralli et al., 2018b.

Identifying the Technology Options for Increasing PNT Robustness and Resilience

In the remainder of this chapter, we provide a description of options for enhancements and alternatives that could strengthen the PNT ecosystem in response to threats to GPS. Reflecting the differences in the desired levels of precision or accuracy of different critical infrastructure sectors discussed above, some options are relevant to some sectors but not others.

Also, some of these enhancements are already a part of the ecosystem relatively broadly; thus, their potential for strengthening the PNT ecosystem would consist of more broadly using those enhancements across applications to better make use of already available capabilities. In other cases, options and technologies have been designed for specific applications that are qualitatively different from serving national PNT needs (e.g., positioning and navigation systems designed to be implemented in single facilities or well-defined areas like ports). For those options, the potential for strengthening national PNT is a question of broader diffusion and adoption, but where the designed end point might not be full national coverage.

To organize this discussion, we have placed the options into six broad categories, summarized in Table 5.6.

- The options in the first broad category are **wireless PNT signals,** systems that provide dedicated RF signals to user equipment, locally computing a PNT solution based on the known location of either space-based or terrestrial transmitters.
- In the second category are technologies that use **RF "signals of opportunity"**—signals that are broadcast for reasons *other than navigation or time transfer* but that can be used to find an estimated location of the receiver by triangulation and,

Table 5.6
Complementary PNT Options

Class of Alternatives	Example Alternatives	Common Characteristics of All Alternatives in Class
Wireless PNT signals	• Multi-GNSS constellations • eLoran • Satelles STL • NextNav • Locata • Pseudolites	• RF signals from transmitters of known location • User computes time and location from pseudorange • Maintains user anonymity
Signals of opportunity	• WiFi/WLAN • LTE/4G Cellular • 5G Cellular • MF DGNSS • AIS	• Existing RF signals not primarily intended for navigation • User computes location from triangulation and ranging • User may or may not get time from signal
Wireless time signals	• WWVB • DCF77	• Usable by remote users • Maintains user anonymity • Provides time only
Wired time signals	• NTP • PTP • Optical fiber • White Rabbit	• Variations on two-way communications to measure time-transfer delays • Provides time only
User equipment based	• Holdover clock • Chip-scale atomic clock • Inertial navigation system • Simultaneous location and mapping (SLAM)	• No new signals • Relies only on modifications to user equipment • Typically sensors and technologies dissimilar to GNSS • May maintain user anonymity
PNT resilience technologies	• Nulling antennas • Direction finders • Jamming or signal degradation detection systems	• Enables continued use of GPS signals and/or existing user equipment • Maintains nominal PNT performance of benign environment

therefore, can be used for navigation purposes. Some signals of opportunity may also include time information.

- The third category covers **RF timing signals** that are only intended to provide time; because of the design of their transmitter network, such signals cannot be used to compute position.
- The fourth captures **wired time signals**, in which information needed to access or calculate precise timing are exchanged on wired networks (e.g., terrestrial com-

munication, fiber-optic, or other dedicated networks). These methods provide time transfer only and have no application to positioning or navigation.

• The fifth includes all options that involve only **user PNT-related equipment** and do not involve RF signals. Options here include methods to "hold over" PNT (initially calibrated against GPS or other sources) and continue activities using the holdover technology through a GPS disruption. Holdover clocks and INS are key examples of such technologies.

• The final category we consider are **PNT resilience technology options;** such options are not new complementary PNT alternatives—rather, they are measures intended to enable the continued use of GPS user equipment, despite the existence of threats to their use. These options should also be considered because they might affect the need for, or choice of, a complementary PNT system.

The divisions between categories are not sharp and the alternatives could cross categories. For example, a dedicated wireless PNT or timing signal from one transmitter or a few transmitters could be used in combination with other less-structured RF signals of opportunity for improved overall results (see Johnson and Swaszek, 2014). Wired time signals could improve the synchronization of transmitters **and** lead to a downstream improvement in wireless PNT performance. Also, besides the variation in capability, accuracy, and precision provided by different alternative sources of PNT and the different contributions of potential resilience or other technologies, many other factors could influence the acceptability of some options. For example, one key feature of wireless signals is that users typically can use these signals anonymously. That is, they do not have to share their derived PNT with any other user or transmitter. This is important for user privacy and is likely to be an important factor in user adoption.

In the remainder of this section, we describe the range of options summarized in the second column of Table 5.4 and discuss their variation in capabilities and costs. For readers interested in greater technical detail, see Appendix C.

Wireless PNT Signals

Because of the attractiveness of being able to fix a position using a single user receiver or access timing information without the need to connect to physical infrastructure, a number of technology options—including those designed for wide areas and systems intended for implementation in smaller areas—are based on use of wireless signals. Many of these methods use four independent signals to compute three-dimensional position and time; fewer signals could be used if position or altitude is considered to be known.

Multi-GNSS Receivers

Even for devices that are primarily designed around using GPS, it is advantageous to also make use of other existing (and potential future) GNSS signals as well. In addition to GPS, GLONASS, Galileo, BeiDou, Quasi-Zenith Satellite System (QZSS), and

Indian Regional Navigation Satellite System (IRNSS), all could be sources of additional satellite-based PNT data. By using a larger number of signals, PNT accuracy can be improved and performance issues such as drop-offs in performance when individual satellite signals are blocked by geography or buildings (discussed in Chapter Four) can be reduced. Multiple redundant signals can also make it possible for receivers to compare the answers they get using different systems, thus making it easier to monitor for inconsistency and threats like spoofing attacks. If all four major satellite constellations could be processed on one chip, then well over one hundred signals would be available globally.

Because of these advantages, many GPS receivers already use two or more GNSS signals. A 2016 survey of GPS receivers finds that nearly 65 percent of receivers exploit at least two constellations, with GPS and GLONASS being the most common among the various possible combinations (European GNSS Agency, 2016). Most smartphones built in the last several years already support this capability. Smartphones that also receive Galileo and BeiDou signals have been released (European GNSS Agency, 2018) and will likely become common in the next few years.

If GPS signals are degraded because of some event specific to the GPS constellation or because of spoofing that mimics the unique structure of the GPS signal, then multi-GNSS receivers may maintain good PNT, with only a slight degradation of performance. However, in the case of deliberate noise jamming, an attacker could likely jam all GNSS signals because of their shared frequency band and weak signal levels.

Redundant satellite GNSS systems can significantly reduce the risk of *spoofing* attacks, given the ability to cross-check information from different systems (or, conversely, the requirement to spoof multiple systems simultaneously). Although using multiple constellations does not fully mitigate the risk of *jamming*, it does address the majority of GPS-system specific threats (specifically, the large-scale successful cyberattack on GPS). Then again, the reliance of states on the functionality of their own GNSS systems also reduces the likelihood those same states would stage *large-scale nuclear or other attacks* in space given the likely collateral damage to their own systems. *Space weather* would remain a concern because it would affect all space-based systems.

Because multiple GNSS constellations already exist and the technology needed to use their signals is already being adopted and proliferated into consumer-level technologies, this technology represents an existing and expanding component of the PNT ecosystem. As such, government action is not needed to promote development or adoption, because market forces and the technical advantage provided by adoption are already driving the processes.

Both GPS and other GNSS constellations are now broadcasting on multiple frequencies, and receivers which take advantage of dual-frequency multiconstellation (DFMC) GNSS are now becoming commonly available, including in smartphones. This modernized DFMC GNSS will deliver significantly better performance than unaugmented legacy GPS.

eLoran

Loran-C was a timing and radionavigation service using high-power signals from terrestrial antennas in the 90–110 kilohertz (kHz) band. It was intended to provide positioning accurate to within about 0.25 nautical miles or 460 meters (Narins, 2014). That level of accuracy was useful to mariners, although it was insufficient for harbor navigation. Still, since the error in repeatability of the position calculation and the relative location with respect to nearby users could be several times better, it was useful for the relative navigation and safety of ships (USCG, 1992). eLoran was designed to be an improved version (providing better positioning accuracy) that would make use of the same transmitter sites and much of the existing Loran-C infrastructure (International Loran Association, 2007; John A. Volpe National Transportation Systems Center, 2009).[1]

The signals used for eLoran have some advantages, notably the ability to propagate without suffering from the line-of-sight limitations that are typical of most other signals designed for PNT. As a result, eLoran signals can penetrate into indoor environments. Signals from eLoran transmission sites can be used for timing by computing the time difference of arrival for any pair of signals.[2] As with other PNT systems, accuracy and performance is improved when more signals from separate transmitters can be used. The extent that an eLoran signals could provide national coverage would depend on the number of sites at which transmission towers were installed.

Even without the benefit of differential corrections, eLoran is expected to provide time transfer with better than one µs accuracy. With a network of reference receivers to provide differential corrections, time transfer accuracy would be better than 100 nanoseconds (ns) (Offermans, Bartlett, and Schue, 2017; UrsaNav Inc. and Harris Corporation, 2017). Differential corrections should allow positioning accuracy of about 20 meters, and one model suggests that accuracy better than ten meters may be possible (Safar, Vejrazka, and Williams, 2011).

eLoran has been deployed largely in other countries. The General Lighthouse Authorities of the United Kingdom and Ireland (GLA) pursued an implementation of an eLoran network with the hoped-for cooperation of other European nations. GLA found that maritime navigation by eLoran within 30–50 meter accuracy was possible (Ward et al., 2015). However, the governments of France, Norway, Germany, and

---

[1]   In 2007, the DoT and DHS jointly recommended that the national backup for GPS should be eLoran (Parkinson et al., 2009). However, in February 2010, the USCG ceased transmitting most Loran-C signals, and in August 2010 U.S. stations that had operated in concert with Canadian stations also ceased transmitting. In all, twenty-nine U.S. and Canadian stations were decommissioned (USCG Navigation Center, 2011). The USCG began dismantling former Loran-C stations until 2014, when the Howard Coble Coast Guard and Maritime Transportation Act of 2014, further extended by the Coast Guard Authorization Act of 2015, directed the USCG to cease dismantling the towers until it could make a determination that the Loran-C infrastructure was no longer needed as a backup for GPS.

[2]   Synchronization is maintained by three atomic clocks at every transmitting station. The redundant atomic clocks improve overall timing accuracy and improve holdover duration in the event of a GPS outage.

Denmark decided that the benefits of the eLoran network did not justify the cost and shut down their transmitters, ultimately leading the GLA to also abandon the project (Saul, 2016; Proctor interview, 2018). Saudi Arabia has upgraded its five Loran stations to be eLoran-capable. India is also reportedly interested in upgrading its six Loran stations and perhaps expanding the network (Narins, 2014). South Korea is pursuing an eLoran system for use in port entry, motivated by concerns over North Korean GPS jamming (Seo interview, 2018). South Korea is currently testing the capabilities and has not reached initial operational capability; the aim for final operational capability (FOC) is 2019.

Unlike with alternative GNSS signals, where receiver technology is being integrated into devices broadly, some user systems would still require significant technology development to integrate eLoran. Currently, receivers are not small enough to be integrated into smartphones, although miniaturization may be possible (Reelektronika, 2017). Current technology would therefore only be applicable to less size-sensitive applications (e.g., some critical infrastructure timing needs).

As a result, for eLoran to make a broad contribution to reducing the risk from GPS disruption or corruption, significant government investment and adoption promotion would be required. Transmitter sites would have to be put in place and maintained; also, while eLoran would not be exposed to space-based threats to the extent of satellite systems, eLoran *would* be subject to a range of ground-based natural and adversarial threats. Implementations with fewer eLoran towers (e.g., implementations that use the system for timing rather than positioning could require as few as four sites) would minimize infrastructure costs, but such implementations would be subject to greater risk from ground-based threats because there would a smaller number of potential targets. Given the need for specific receiver equipment, many users would need to acquire new receivers. The likelihood of users making that investment (in contrast to the multi-GNSS case above where market forces are driving integration) is not clear. Current costs of receivers are in range of thousands of dollars, but even large percentage reductions in that cost with increases in production would still mean that cost could be a barrier for some classes of users to adopt.

Because this technology is not implemented and there are not clear market drivers for putting the infrastructure in place or driving adoption in use once built, significant government action (and associated costs) would be required for eLoran to make a contribution to national PNT robustness and resilience.

Satelles Satellite Time and Location

Iridium has had a constellation of 72 communication satellites (66 operational satellites and six spares, in six orbital planes) in low earth orbit (LEO) (Graham, 2018). While this constellation has the primary mission of providing commercial mobile voice and data communications globally, it has recently been utilized to transmit PNT signals, providing a PNT solution from LEO that is independent of GPS in collaboration with

Satelles and other firms (Reuters, 2016; GPS World Staff, 2017; Spectracom, 2018). STL is a subscription-based product where the company provides user equipment and service support to customers (Iridium, 2016).

Iridium transmits in a different frequency from GPS signals, although not in a radically different band, and uses a much more focused transmission geometry. As a result, the power of the signal at ground level is much stronger than GPS and, therefore, less susceptible to jamming. However, its positioning and timing accuracy is lower fidelity: STL positioning accuracy is advertised as 30–50 meters and timing is 200 ns compared with typical GPS performance of about five meters positioning accuracy and about 20 ns timing accuracy (Spectracom, 2018). As a space-based system however, STL would still be exposed to some of the same risks as GPS with respect to natural, cyber, and national security-driven threats.

Because this system already exists and is available in the market, government investment in infrastructure is not required. The contribution of this service to national PNT resilience is therefore constrained by user adoption, which has associated costs given the subscription model for the service. The current extent of user adoption could not be assessed: Although firm communications claim that "many from industry and government are already using this service to achieve a more robust PNT solution," specific information identifying the number or identity of those clients is not publicly available (Spectracom, 2018). As with eLoran, current devices are not small enough for integration into size-demanding applications but might be miniaturized with further development. Exact cost estimates of this service have not been publicly released;[3] however, any annual costs associated with the service would be a barrier to adoption over, for example, the use of multi-GNSS satellite signals that are being integrated into many devices as a result of market forces.

Consequently, while potential government costs associated with promoting this system as a part of a more robust PNT ecosystem would be low with respect to *infrastructure*, costs to drive *adoption by users*, which would be required for the system to make a substantial contribution beyond any that have already elected to purchase it on the market, would not be low; depending on exact per-user initial or recurring costs, the costs could be considerable.

NextNav Metropolitan Beacon System

The NextNav Limited Liability Corporation (LLC) has developed a system of terrestrial beacons to provide precise PNT signals to mobile device users in covered areas. Branded as the MBS, it can serve as a complementary PNT source where GPS signals are unreliable (e.g., urban canyons) or provide primary PNT capability where GPS signals are too weak to penetrate (e.g., indoors). The technology provides a signal that is much more powerful than GPS satellite signals at ground level. NextNav claims the

---

[3]   Supporting analysis is provided in the proprietary Appendix G to this report.

system can provide under ten-nanosecond signal timing. Because vertical accuracy based on ranging signals is typically poor for terrestrial systems, NextNav uses differential pressure measurements between sensors in the beacons and user devices to determine altitude. It claims a one to two meter (or "floor level") vertical accuracy for indoor positioning.

The MBS has been installed in several metropolitan areas for demonstration (Gates and Pattabiraman, 2016). NextNav claims five to ten meter horizontal accuracy throughout the metropolitan regions where the system has been installed, which includes a "commercial-grade" deployment in the San Francisco Bay area (over 900 square miles between Marin County and San Jose) and "initial builds" in 39 other metropolitan regions as of October 2013. Beyond these initial deployments, the extent of the potential contribution of such a system would be driven by the scope of deployment. Like other ground-based systems, national coverage would require much broader (and therefore more expensive) infrastructure investment than local implementations. Infrastructure costs associated with system implementation is not publicly available.[4]

In addition to considering national PNT needs given concern about threats to GPS, efforts have also focused on the needs of emergency responders and 911 emergency call systems because of the transition of many individuals from wired to mobile phones. While legacy 911 systems could readily link a call to a specific address, aiding dispatch and response, calls to 911 from mobile phones are more of a challenge. To solve this problem, the enhanced 911 (E911) initiative is focused on modernizing 911 systems to use mobile positioning data to locate callers to provide similar aid to response. The ability of MBS to provide positioning in urban environments and indoors, with floor-level accuracy, significantly benefits E911's goal of automatically providing the location of callers to 911 dispatchers. Efforts related to E911 have been a part of the FirstNet program (a program addressing first responder communication and other needs with AT&T as the prime contractor), and AT&T and NextNav have signed a term sheet in support of meeting FirstNet's enhanced location with z-axis capabilities requirement. At the time of this writing, the firms report that they expect to have a definitive agreement signed in the coming months in support of this effort and to roll out the service commercially.[5] To the extent that a decision is made to implement this system under FirstNet in populous areas of the country to address E911 capability and first responder positioning within urban built environments, that same investment will incidentally provide supplemental PNT coverage to other users in those areas as well.

As is the case for multi-GNSS signals, the technology for integrating the ability to receive and use these signals is being developed and deployed at no cost or minimal

---

[4]   Supporting analysis is provided in the proprietary Appendix G to this report.

[5]   Email from Ganesh Pattabiraman, chief executive officer, NextNav, December 7, 2018.

cost to consumers.[6] Mobile processors capable of supporting the MBS are being integrated into commercially available mobile devices today, and the rapid refresh rate of such devices will mean adoption by consumers within near-term time periods.

Although government costs associated with expanding this option beyond whatever areas FirstNet will cover would be considerable, that decision will lie in the future after the costs and performance of the FirstNet system can be assessed. Certainly, with the ability to use these signals already being integrated into user devices, adoption promotion costs would appear to be low.

Locata

The Locata Corporation has developed a system of terrestrial beacons to provide PNT signals to dedicated receivers in a localized area. The system reportedly can provide centimeter-level positioning precision and under one ns timing synchronization between transmitters without the use of atomic clocks. Locata uses a proprietary signal in the same band as Wi-Fi transmitters and therefore performs similarly to Wi-Fi with respect to signal obstruction and interference.[7] The system consists of transmitting beacons. The original intent was to develop a system to provide positioning in locations where GPS signals could be obstructed or degraded, such as deep valleys, opencut mines, forested areas, and urban and indoor environments (Rizos, Gambale, and Lilly, 2013). The system has been used in warehouses, mines (Rizos, Gambale, and Lilly, 2013), government facilities (Craig et al., 2012), airports, and sea ports to back up GPS and/or improve available precision, especially for logistics applications.

Because it operates in a different frequency band than GPS, jamming the system would require different jammers than attacking GPS alone. As a terrestrial system, it would not be exposed to the same risks as space-based systems, but it would be affected by ground-based threats and hazards. As with other locally installed systems, the breadth of the potential contribution of this system to national PNT robustness and resilience (versus the system's design to cover key sites where backup PNT was needed) would depend on how widely it was implemented, and the infrastructure costs would scale with the breadth of that implementation.

While companies like Leica have created integrated user devices that act as receivers for both Locata and GPS, there is no standardized Locata microprocessor that has been integrated with consumer-level user devices, such as what is happening with multi-GNSS chips and the NextNav MBS system. Such development may be possible, but at present, deployment of a LocataNet generally requires users to purchase all parts of the

---

[6]   For the E911 application, low barrier to individual adoption in smartphones is critical to achieve the intended life safety goals.

[7]   Locata indicates in their on-line FAQ that their system will not interfere with Wi-Fi systems operating with it if properly configured (Locata, undated).

system.[8] This would represent an important barrier to adoption, and adoption promotion could therefore become a significant government cost if this system was intended to become a significant contributor to national PNT robustness and resilience.

Use of Locata in Aviation

Locata has demonstrated high vertical precision (0.15 m) in tracking high-altitude aircraft over an Air Force test range at White Sands (Craig et al., 2012). To help assess whether Locata could support the precision landing applications listed in Table 5.4, the HSOAC team modeled the performance of a ten-transmitter Locata installation similar to the one at White Sands.

Figure 5.1 shows the vertical dilution of precision (VDOP) one would predict at various altitudes over the installation, given the transmitter geometry. The transmitters are located on hills with altitudes varying over a range of 3,000 ft, and horizontal locations indicated by black 'x' marks. One would expect that at altitudes over 20,000 feet above mean sea level (MSL), this LocataNet would have a VDOP between two and three over most of the test range between the transmitter locations, and indeed, this is confirmed by the test performance reported in Craig et al. (2012), in which an aircraft at 25,000 feet measured VDOP at 2.7 and vertical root mean square (RMS) error at 15 cm. If the aircraft were to descend to lower altitudes, however, the VDOP would become worse because the geometry of the terrestrial transmitters would be less favorable for determining vertical position. Close to the ground, the VDOP would often be over 25 so the vertical RMS error would be over 1.5 m and the 95-percent error would be over 3 m. This is too much vertical error for a CAT II or CAT III precision landing, but it might be adequate for a CAT I precision landing.

Pseudolites

Pseudolites refers to terrestrial transmitters that broadcast signals compatible with existing GNSS user equipment on the same carrier frequency using a signal structure that is the same as GPS or any other GNSS (Raquet, 2013). A key motivation for such systems is indoor PNT where space-based signals are not available. Pseudolites can also have much higher signal strength than normal GNSS signals and thus much greater resistance to jamming. Because of the terrestrial geometry of pseudolite transmitters, horizontal positioning using such systems would be about as accurate as space-based PNT, although vertical positioning would be inferior. Several approaches have been proposed for using these technologies, reflecting the practical issues associated with data provided by a ground-based, immobile transmitter being interpreted by devices designed to use data transmitted by mobile satellites. At least some pseudolite concepts might potentially be implemented at relatively low cost using software-defined radios and other commercially available equipment. However, there are concerns, including the potential that pseudolites might create interference for space-based GNSS signals.

---

[8]   More detailed cost information is included in the proprietary appendix to this report.

Figure 5.1
Predicted Vertical Dilution of Precision from LocataNet at White Sands

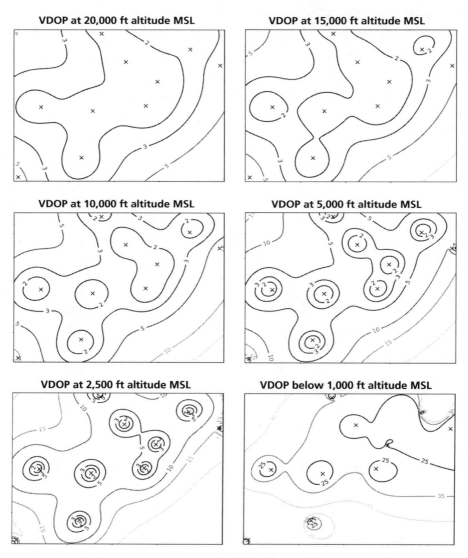

The associated challenges mean that no pseudolite architecture based on the GPS/GNSS carrier frequency has been commercialized. Indeed, it would be illegal to broadcast signals that interfere with GPS. Any pseudolite architecture would have the difficult task of proving it is compatible with GPS before it could be entertained as a viable backup option.

Signals of Opportunity

Signals of opportunity (SoOPs) are defined as RF signals that are not intended for navigation (Raquet, 2013). These can include such things as AM/FM radio, digital TV, Wi-Fi (IEEE 802.11), Bluetooth, RFID, and cellular telephony. Advantages of SoOPs are (1) they are ubiquitous; (2) many are high power compared with GPS and therefore would be relatively difficult to jam; (3) many can penetrate buildings or are available in indoor environments; and (4) common user devices like cell phones already receive many of these sorts of signals. Disadvantages of using such signals for PNT are (1) since they were not designed for PNT, resulting PNT accuracy is likely to be worse than GPS; (2) positioning is possible only if transmitter locations are known; and (3) such signals will not be available everywhere.

Apple and Google, as well as third-party providers such as Skyhook Wireless, maintain maps of transmitter locations to aid in positioning. These "navigation clouds" can be maintained in large part by crowdsourced data sent back from smartphones and tablets to the operating system providers (Fleishman, 2011; Huang and Apple Inc., 2014).

Software-defined radios (SDRs) could generalize such a concept because they could potentially make the fullest use of transmitters on the RF spectrum. Data analytical methods could infer and continually improve knowledge about transmitter locations, which could be available to users if GPS becomes unavailable. This might present user privacy concerns, but it is technically plausible.

Below, we consider three main SoOPs—Wi-Fi, LTE (4G) cellular service, and 5G cellular service—as most relevant for land-based PNT. Another set of SoOPs—medium-frequency differential GNSS broadcasts and AIS broadcasts—has been proposed in the context of maritime navigation.

Wi-Fi/Wireless Local Area Network

Wireless local area networks (WLANs) are commonly based on Wi-Fi signals. Such WLANs are ubiquitous; many homes and commercial establishments have them. Our smartphones detect them, even when they are secure networks that cannot otherwise be accessed. Ranges of such signals range from a few tens of meters for simple home installations to perhaps many times more for more powerful and extended networks, as one might find at a hotel or office. Koivisto et al. (2017) report that ranging accurate to three to four meters is possible. The accuracy of Wi-Fi positioning on typical user devices has been reported to be up to 2.5 meters (Empson, 2013) but perhaps more commonly 10–20 meters (Skyhook Wireless, undated).

Since Wi-Fi signals are already available to a vast number of user devices, the main challenge for employing this approach is maintaining the database of transmitter locations. As mentioned, Apple and Google maintain "navigational clouds" of this type by using crowdsourced information from smartphones (Fleishman, 2011), thus providing users with position and navigation knowledge that is comparable to GPS in environments where multiple Wi-Fi and WLAN signals are simultaneously detected. While this relies on the prior location information for the Wi-Fi transmitters, which is established usually through GPS, that location is unlikely to change.

The Wi-Fi and WLAN alternatives can generally provide positioning and navigation comparable to GPS with little or no modification to common user equipment. This is particularly attractive for cell phones, which by far represent the largest class of receivers. Moreover, the signals are generally ubiquitous in urban and suburban areas, even indoors. Deployment is generally not dense enough in rural areas for positioning. However, it is unclear whether Wi-Fi can support time transfer. For the present discussion, we assume that it does not. As terrestrial systems, these technologies have different threat and hazard exposure than space-based systems, and their diffuse nature means that they would be highly difficult (and even impossible) for an attacker to disable (outside of scenarios where there are massive attacks on the power or communications grid that would have consequences separate from PNT loss). Of course, they could be jammed locally but not easily over large areas. As preexisting infrastructure that is built by individuals and sustained by communications and other providers as part of commercial activities, these elements of the PNT ecosystem already exist at no cost to government and do not require government investment going forward.

LTE (4G) Cellular

LTE is a specific implementation of 4G cellular telephony that is in common use in U.S. markets. According to International Telecommunications Union standards, LTE base stations should remain synchronized with Universal Time Coordinated (UTC) within ±1.5 μs (Goode, 2018). This is easily maintained when base stations have access to GPS signals. This degree of synchronization allows users to infer location data from LTE signals. As is the case for GPS signals, user devices like cell phones determine their position based on time differences in arrival of signals from different base stations. LTE is reported as being able to typically provide positioning accuracy of several tens of meters, possibly also using measurements of signal strength (Koivisto et al., 2017). The intrinsic time uncertainties in base stations dominate the quality of time transfer, with users able to maintain time to nearly ±1.5 μs of UTC, as long as the base stations maintain time to this accuracy.

As a terrestrial alternative to GPS, 4G-LTE is likely to be useful for PNT against a wide range of threats as long as base station synchronization remains adequate for time-based positioning. This is likely to be true for outages even as long as months. Only for very long-term outages might positioning and navigation eventually degrade to the point of becoming nonuseful. Moreover, these signals are ubiquitous, even more so than Wi-Fi because they extend into many (though not all) rural areas. As a result of their ongoing business operations, providers of wireless communications also have an incentive to maintain service (e.g., to repair towers damaged in ground-based incidents) and, therefore, PNT functionality without government expenditure. There are also not separate consumer costs associated with access to PNT from this source separate from their communications subscription plan with a provider.

5G Cellular

Although the standards for 5G cellular telephony are not fully established, some of its characteristics can be anticipated. Because of the nature of the technology associated with 5G, the expectation is that there will be a much higher density of transmitter nodes to connect with user devices (Roessler, 2017; Koivisto et al., 2017). This is expected to allow devices to use 5G signals to determine their location within one meter or less. This is considerably better than what is now possible with GNSS alone. Moreover, there will be a significant incentive for users and providers to implement such capability. Whether the computations to localize and track occur in user or provider equipment appears to be a choice that has not yet been made.[9] Much like 4G-LTE, the synchronization requirement for a 5G network is expected to be ±1.5 µs across the entire network, but the synchronization among adjacent nodes might be considerably better (Goode, 2018). Hence, it is likely that positioning and navigation based on 5G can be sustained for long periods without GPS or GNSS. The quality of synchronization will probably also degrade only slowly following a GPS/GNSS outage, although the absolute accuracy of time transfer would probably degrade at the rate implied by the quality of any holdover clocks in the 5G network.

Because 5G implementation is already in progress, the United States will probably see the first major deployments of 5G infrastructure over the next few years. This technology could meet the PNT needs of many critical infrastructure sectors, with few implications for acquisition other than those already expected under consumer market forces. The only question may be whether 5G will be available in remote areas, where providers may not have an incentive to invest in the cost of dense transmitter networks.

As a terrestrial alternative to GPS, 5G is relevant across a wide range of threats, providing positioning and navigation superior to GNSS and GPS as long as base station synchronization remains adequate and timing synchronization can be passed to the 5G cells. Additionally, the existing 4G-LTE network described above will continue to function for months even if the 5G cells degrade. As with 4G-LTE, implementation of these technologies are driven by market forces (and largely distinct from PNT as a specific function); as a result, there is little need for government investment or expenditure for either infrastructure or user adoption.

Medium Frequency (MF) Differential GNSS (DGNSS) and Other Signals

Johnson and Swaszek (2014) proposed using MF-DGNSS broadcasts, normally used to transmit correction and integrity information for the GNSS, in a ranging mode, possibly together with other existing SoOPs such as AIS broadcasts. Their feasibility

---

[9] The choice will have implications for computational demands on user equipment and also issues for user privacy.

study looked at navigation in the North Sea using MF-DGNSS, AIS, and possibly a single eLoran transmitter. Using MF-DGNSS alone, the achievable accuracy was better than ten meters in most of the North Sea area during daytime, but about a factor of ten worse at night. Using AIS signals alone was able to give ten-meter position accuracy near the coast, where AIS transmitters were within line of sight. The combination of all three signals gave the best performance, with better than ten-meter accuracy near the coast and at least 100-meter accuracy in the center of the North Sea.

Wireless Time Signals

Beyond time information being provided as part of an overall PNT system, there are also dedicated wireless systems intended for timing alone.

WWVB

WWVB is a radio timing signal broadcast by the National Institute of Standards and Technology (NIST) from Fort Collins, Colorado.[10] The signal can be received across most of North America, although reception varies with time of day and is best at night. Apart from the propagation delay of the radio signal, WWVB has an inherent time accuracy of about 100 μs. The radio signal delay could introduce an offset of up to 15 milliseconds (ms) in the CONUS (DoD, DHS, and DoT, 2017), although this can be corrected in many cases for the path and reduced to 100 microseconds (Lombardi, 2003, p. 26). A key issue for WWVB users is that electromagnetic interference (EMI) from a variety of sources can prevent signal reception (Lowe et al., 2011).[11] The orientation of receiver antennas can also be a source of loss. Steps have been taken by NIST in response to these issues. Receivers that exploit the modified signal are now commercially available, including the radio-controlled clocks and watches that are available in the market.

The most notable advantage of WWVB for time transfer is that the infrastructure already exists. There are no implications for new national infrastructure to sustain the timing signal, unless there is a need to make this signal available to users in Alaska and Hawaii. There is already a market for radio-controlled clock devices. WWVB antennas can be small enough for devices as small as smartphones. However, the relatively low accuracy without path correction and continuing potential for EMI might make the resulting performance inadequate for some applications, and the unpredictable orientation of antennas in small devices will be another source of signal loss. It is

---

[10] WWVB should not be confused with NIST's WWV and WWVH signals, which are high-frequency broadcasts from Fort Collins and Hawaii, respectively. These signals provide less accurate time transfer than WWVB but can be received at much longer ranges.

[11] Machinery and electronics are a major source of interference, but there are many other sources of man-made and natural EMI, including distant lightning, multipath interference, diurnally dependent atmospheric losses, shadowing and obstruction by geography, and on the East Coast, the similar "Rugby" timing signal broadcast from the United Kingdom.

possible that even with an upgraded signal, some users will be able to acquire the time only in nighttime conditions. This might be sufficient for users that do not require high-accuracy time, because almost any clock could provide adequate holdover capability for a day.

DCF77

As an example of another wireless time signal comparable to WWVB, the DCF77 protocol is an international time protocol accurate to 100 μs, designed and broadcast by the Physikalisch-Technische Bundensanstalt; it is used to disseminate German national legal time and is used as a time source by most radio clocks in Europe. The 77.5 kHZ radio signal can be received indoors. The protocol is well suited to power and industrial applications because it can be distributed to devices as a 24 VDC level-shift signal (Kennedy, 2011).

Wired Time Signals

The majority of PNT alternatives discussed to this point are one-way methods because receivers do not need to send back any information to transmitters about their own PNT solution. In this section on wired alternatives to GPS, all methods are two-way because wired communications routes are often switched and routed and are subject to delays and latency of network traffic. Time to send a signal from one device to another is not a direct or even approximate measurement of distance. Hence, all wired alternatives to GPS may provide time transfer but are not candidates for positioning and navigation. Compared with wireless systems for timing, wired applications have the advantage of being entirely free of risks from RF interference and can potentially be used in indoor and underground applications where GPS or other RF signals are unavailable.

Network Time Protocol (NTP)

Developed in the 1980s, NTP is commonly used to distribute time from a computer server to its clients on a local area network. By design, it minimizes the use of computer resources to potentially provide time to up to many thousands of clients. It is built into commonly used operating systems like Windows. A frequent application is to provide time to the integer second. In typical situations, it provides time to an accuracy of only about one ms (e.g., Microsoft, 2016). A recent effort to understand the fundamental limits of NTP achieved 10–20 microsecond accuracy by carefully eliminating time asymmetries (Novick and Lombardi, 2015). However, such performance cannot be expected in most existing networks.

Key advantages for NTP are that it requires no specialized hardware and is already available to most computer users, it can function entirely independent of GPS, and it can be used where RF signals are unavailable. Key negatives are that it is only available to users on a computer network and its accuracy is comparatively low. Notably, NTP appears to be good enough for the timing needs of the wired communications sector itself (i.e., good enough for wired communication, including conversational video, conversational voice, and even mission-critical delay-sensitive signaling, given fixed routing to avoid variable packet delays in an internet protocol–based network). However, NTP is inadequate to support the synchronization needs of some wireless applications that could be connected to the wired network. But its ubiquity means that government involvement is not required for broader use as a supplement to GPS-provided timing where appropriate.

Precision Time Protocol (PTP)

PTP, also known as IEEE 1588, was introduced in 2002 and a revised standard, sometimes called PTP v2, was issued in 2008. It relies on the same basic algorithm as NTP, but its implementation is more complex (Klecka, 2015). It supports networks of networks to minimize asymmetries and allow the best possible synchronization among peers in a network, and it uses specialized hardware to improve the quality of time stamping. Because PTP uses multiple different types of clocks[12] for synchronizing time over a network, it is scalable and can achieve submicrosecond accuracy (Eidson, 2005). It is already used by many telecommunications networks as well as a wide range of industrial, commercial, and scientific users. Like NTP, it could function almost entirely independent of GPS. If the Grandmaster clock is a high-fidelity source of UTC, such as a clock at NIST, then the network could sustain submicrosecond accuracy independent of GPS. As with other existing infrastructures, acquisition costs would largely be borne by users. In this context, users could be the end user or perhaps telecommunications providers that maintain and operate nodes in the backbone network.

Precision Time Protocol over Optical Fiber

A recent study sponsored by DHS sought to test the limits of PTP over long distances between telecommunications nodes connected by optical fiber (Weiss et al., 2016). The key feature of this experiment was the ability to calibrate out the time asymmetry because there is a high-quality atomic clock at each end of node: an atomic clock at NIST in Boulder, Colorado, and the U.S. Naval Observatory's (USNO's) Alternate Master Clock for GPS at Schriever Air Force Base in El Paso County, Colorado (Cosart and Weiss, 2017). PTP over optical fiber could be a good method of backing up GPS time for any sites connected to optical fiber, which would typically include transmitters in a wireless cellular network. The demonstration showed that this method can support time transfer over significant distances, assuming that the system was put into place when GNSS was available and its functioning is maintained through a period of outage (i.e., if the system was turned off and back on during an outage, it would

---

[12] Such clocks include the Grandmaster that serves as "truth" for the network; ordinary clocks that only receive time; boundary clocks that can serve time to subdomains; and transparent clocks that try to pass time to subdomains with as little effect as possible (adding only a time correction for network delay) but otherwise are not serving time.

be difficult to recover operations). When operating normally, the method can support time transfer with accuracy that exceeds the needs of almost all users, but its application would be limited to the subset of users with resources (e.g., existing atomic clocks) required for its use. As a result, it could not provide direct time transfer to the majority of end users, such as mobile smartphones or bank automated teller machines (ATMs). However, it could be used to complement the main backup for such users. For example, if the main GPS backup for such users was cellular telephony, PTP over optical fiber could keep the major telecom nodes synchronized.

White Rabbit

White Rabbit technology development began at the European Organization for Nuclear Research (CERN) in 2008 to provide subnanosecond time transfer for particle physics applications over distances of about ten km, which is farther than the distance between nodes in a typical PTP subdomain (Serrano et al., 2013). This technology now has been demonstrated and commercialized. By design, it is intended as a high-accuracy extension of IEEE 1588, and it may soon be officially incorporated into the third release of the PTP. It is backward compatible with earlier releases of IEEE 1588. White Rabbit provides this precision by using multiple synchronized clocks in the network and by how the network does time stamping of the messages that are exchanged. Resulting accuracy of time transfer depends on how many levels of nodes are used, but even the earliest demonstrations in 2013 all consistently performed with better than one ns accuracy, often several times better, with accuracy to within 200 picoseconds. A recent demonstration using upgraded White Rabbit hardware switches over a 500-km link constructed of four levels, each 125 km apart, still exceeded GPS accuracy (Kaur et al., 2017a, 2017b).

White Rabbit master units and switches are sold commercially by Seven Solutions.[13] These units are rackable in standard 1U 19-inch racks. White Rabbit technology was developed by an international collaboration using open hardware and open software principles. The nonproprietary nature of this technology could facilitate evolutionary improvements or expansion into broader markets. The master unit's source of time could be GPS, but it could also be any other high-quality source like NIST's UTC atomic clock standard. If the network is independent of GPS, then its performance will be unaffected by any GPS outage.

White Rabbit can support time transfer with accuracy that substantially exceeds the needs of almost all users; it is better than GPS. Therefore, this method is a strong candidate for backing up GPS time transfer for users that require atomic clock accuracy and for serving as a "national backbone" for time so that secondary users, such as cellular networks, can perform to the limits of its own subdomain without suffering additional inaccuracies of its own master clock. Less accurate methods, like ordi-

---

[13] It is not clear whether Seven Solutions is yet selling the upgraded hardware that was used in the 500-km demonstration.

nary PTP, could provide timing to the vast number of other users, like mobile and cellular users.

User Equipment Alternatives to GPS

Wireless or wired systems, whether specifically for PNT or for utilizing SoOPs, focus on the national PNT ecosystem at the macroscale and are options for strengthening robustness and resilience either for significant numbers of users or for the nation overall. There are also technologies and tools that individual users can adopt on their own or integrate into their own infrastructures or systems that make *their* access to PNT more robust or resilient. Depending on the scope of the adoption of such alternatives, the nation overall might become more resilient, even if benefits accrue from the bottom up for individual users. The analysis in Chapter Four about jamming alluded to examples of these technologies (e.g., backup clocks in financial sector firms).

Holdover Clocks

When GPS is degraded, holdover clocks can provide time when GPS goes out. Of course, integrating holdover clocks into user equipment is not new; virtually all user equipment has clocks capable of holdover for brief GPS outages, but the accuracy of such clocks relative to UTC tends to degrade with time. Until recently, atomic clocks that could maintain better than one μs accuracy after 24 hours were the size of electronics boxes, cost at least a few thousand dollars, and used at least tens of watts of power. The highest fidelity atomic clocks have performance orders of magnitude better, but the best performance is achieved only in laboratory-scale systems (Shkel, 2011).

However, beginning in the early 2000s, the Defense Advanced Research Projects Agency (DARPA) sought to develop an atomic clock capable of providing one μs accuracy after 24 hours in a device of one cm^3 size, using less than 30 milliwatts (mW) of power. The result was a chip-scale clock commercialized by Symmetricom (now Microsemi) in 2011. This clock uses about 120 mW and has volume of about 17 cm^3, and its accuracy appears to be close to the goal of one μs at 24 hours (Microsemi, 2018). The 2011 cost was about $1,500. The size and performance may be good enough for some critical infrastructure applications.[14] Acquiring and integrating holdover clocks is strictly a matter of user equipment and would be done on a case-by-case basis as need drives adoption. However, to date, it appears that the size and cost of even the smallest such clocks are still too large for the most compact user devices, such as smartphones.

---

[14] DARPA is continuing to develop other precision timing capabilities. In the Quantum-Assisted Sensing and Readout (QuASAR) program, DARPA is trying to make very accurate, laboratory-quality atomic clocks portable for use at field sites. Such capability could potentially improve the holdover performance at important infrastructure sites requiring very accurate time. We do not have information on the current status of this program. One goal of the QuASAR program is to transition the technology into another DARPA program called Spatial, Temporal and Orientation Information in Contested Environments (STOIC), one goal of which is to develop accurate atomic clocks for tactical military applications.

Inertial Navigation System

An INS uses an IMU to measure angular and linear accelerations using gyroscopes and accelerometers. When these accelerations are integrated over time, the INS is able to compute position. INS systems are essentially a holdover capability for position and navigation, assuming an initial calibration using GPS or other precisely located position. INSs are routinely integrated into a wide range of devices, from smartphones to vehicles of all sorts. Other government funded efforts have invested in development of miniature INS devices for a range of purposes (Shkel, 2011), however, performance targets for such devices are not sufficient for long-term holdover positioning and navigation during GPS outage. Even the highest fidelity current INS systems are not suitable for many civilian applications by themselves. Position must be reinitialized at least every few minutes. However, SoOPs or other sources of information might provide these position updates. For example, one study showed that a low-cost, commercial-grade INS aided by cellular CDMA signals could maintain navigation quality better than a tactical-grade INS alone (Morales, Roysdon, and Kassas, 2016).

As was the case for holdover clocks, acquiring INS devices is strictly a matter for user equipment. Developers of user equipment will undoubtedly take advantage of improvements that are cost-effective and useful. Deployment of civilian INS will likely be driven by the need to cope with short interruptions and blockages of GNSS (e.g., indoors) than by concern about large-scale GNSS outages.

Simultaneous Location and Mapping (SLAM)

SLAM is the solution to the problem of an autonomous vehicle moving through an unknown environment, in which the vehicle must build its own map of its surroundings and estimate its position within it as it moves. Examples of existing applications are robots moving through unfamiliar (often indoor) environments (e.g., robotic vacuums) or rovers on Mars.

There are two major aspects to SLAM. The first is the selection of an algorithm that simultaneously estimates parameters of the map, the orientation of the vehicle, and the position of the vehicle within the map. The second is the selection of sensors and signal processing that provides the data for estimation. Since the early 2000s, the algorithmic aspects of SLAM were considered to be fully solved (Durrant-Whyte and Bailey, 2006). The key input data for SLAM are the parameters describing "landmarks" that can be observed by the sensors used for positioning. As long as the sensors correctly *associate* landmarks on sequential views and when landmarks are revisited at later times, position estimates will remain as accurate as the sensor data supports. The most common types of sensor used in terrestrial applications are visual or infrared imagery and light detection and ranging (lidar) (Cadena et al., 2016). Imagery usually provides only angular measures to landmarks while lidar provides range. These can be used independently or in combination. Although SLAM is understood at the basic level, research continues, seeking improved performance depending on nature of the application and environment. A key emerging application will be the use of SLAM in autonomous vehicles. Autonomous vehicles will likely be a market force toward improving capability and cost reduction in many consumer applications. When GPS is available, SLAM can augment performance by developing the local dynamic map where the dynamic elements are other traffic or changes in the map, such as from construction.

SLAM algorithms are not needed if the map is fully known; it is needed only to the extent that there are dynamic elements in the map or if the autonomous vehicle goes off the map. The latter condition might be more common for emergency vehicles. When the map is known, the position simply needs to be updated often enough that an INS can maintain the required accuracy. In an environment without GPS, autonomous vehicles will need to access data in a "navigation cloud" that includes enough information on landmarks to support the sensors that are continually imaging their surroundings. These landmarks would be adjacent buildings, structures, or terrain (or features on these). The main requirement for SLAM to become a broadly applied augmentation to GPS is the establishment of a "navigation cloud" to host the sort of information that can be exploited by vehicle-based sensors. Such data infrastructure may have to be privately developed to support autonomous vehicle markets but could alternatively be provided as a public asset. Depending on the trajectory of the autonomous vehicle market, similar market pressures that have pushed adoption of other positioning and navigation adoption could drive development.

PNT Resilience Technologies

Although the preceding discussion has focused on technologies that directly provide all or a subset of PNT capabilities, a complete view of potential complements or supplements to the GPS system must also include technologies that increase the resilience and robustness of the existing system or address threats directly rather than just those that add additional capabilities to the ecosystem. Of the options to do so, notable examples are the following:

- **Nulling antennas:** Antennas that are designed to making jamming less effective are called nulling antennas, which are designed to admit the desired GPS or GNSS signals with minimal attenuation while canceling out or nulling jamming signals. Options for these antennas range from very inexpensive (e.g., designs that reject signals that are coming from below the horizon) to more sophisticated antennas that specifically reject jamming signals from identified directions. While such antennas are too large to be applicable in small form factor applications like smartphones, they could be used in other applications and are a viable (and comparatively inexpensive) response to jamming threats that would be implemented on a user-by-user basis.

- **Direction finders:** Direction-finding technologies help to detect and, in some cases, geolocate sources of jamming. Such technologies can be handheld devices (e.g., that would be used by law enforcement or others that were responding to jamming incidents) or can be deployed permanently. The ability of these technologies to detect and halt jamming incidents would be proportional to the scale of their deployment and readiness to respond when incidents are suspected.

- **Jamming or signal degradation detection (including "jamming ticketing cameras"):** In contrast to deployed direction finders, if jamming events become sufficiently common, devices could be deployed to respond to the interference in a more automated or systematic way. A limited implementation could involve stationary GPS receivers providing integrity monitoring; that is, because they would be known to be stationary at a specific location, apparent changes in that location would indicate signal degradation or corruption. Communication of apparent incidents of degradation to users would limit the chance the users would rely on incorrect positioning or timing data and could also inform private decisions to invest in backup technologies. If individual incidents of jamming increase (e.g., through the use of privacy jammers), stationary devices to detect the jamming and capture data on the source—as is now done with cameras that capture license tags of vehicles that run red lights—could provide an alternative way to respond. Because such jamming is illegal and has associated fines, this approach could deter the use of such jammers in a reduced cost (or even cost-neutral) way.

Considering System Performance Versus User Fidelity Ranges

This section identifies which of the systems just described would be appropriate for serving which users. The usefulness of the capabilities provided by the specific complementary technologies or alternative PNT options discussed above varies considerably, depending on the user or use case for PNT. This simply reflects very different fidelity ranges for different industries in Tables 5.1 to 5.5. The several orders-of-magnitude difference between the higher and lower fidelity applications as well as the sensitivity of the applications to timing and positioning information within the different contexts of the industries means that different technical solutions are generally relevant to subsets of industries and needs. Table 5.7 summarizes the performance of different options for both positioning and timing based on the information presented in this chapter and the additional technical detail in Appendix C. Some suggested technologies, information on which is not publicly available, are omitted here; however, most such technologies are at a low enough level of development that implementation would not be feasible in the near term.[15]

---

[15] See the proprietary Appendix G of this report.

Table 5.7
Review of Complementary PNT Technology Performance Characteristics

Complementary Technology	Positioning Accuracy	Timing Accuracy
Legacy GPS	5 m	20 ns
GNSS DFMC	30 cm	1 ns
eLoran	50 m, 10–20 m with differential correction	<1 μs, 100 ns with differential correction
Iridium-based	30–50 m, <10 m with differential correction	100 ns
NextNav MBS	5–10 m horizontal, 1–3 m vertical	<100 ns
Locata	<10 cm	0.2 ns
Wi-Fi/WLAN	3 m	N/A
LTE/4G cellular	30 m	1.5 μs at 24 hours
5G cellular	1 m	1.5 μs at 24 hours
DGNSS/other	10 m	100 ns
WWVB	N/A	100 μs
DCF77	N/A	100 μs
NTP	N/A	1 ms
PTP	N/A	1 μs
White Rabbit	N/A	<1 ns
Holdover clock	N/A	1 μs at 24 hours
Inertial navigation	Highly variable	N/A
SLAM	Highly variable	N/A

The highest fidelity demands will mean that timing for those sectors will be on the order of a microsecond in terms of timing while those focused on positioning will demand a centimeter or better level of accuracy. Systems better tailored for more relaxed positioning requirements will require significant augmentation, such as GPS receives through use of differential systems to achieve necessary levels of accuracy. Likewise, stringent timing demands in some applications suggest that any system that cannot immediately provide necessary timing accuracies will require supplemental assistance to ensure that very high precision timing is available to the end users.

Table 5.8 compares proposed system alternatives (that are intended to deliver a timing signal) with reported user needs for timing (Table 5.1). All the timing applications under consideration occur on a power or communications network. Even if it is a wireless communications network, the nodes that need to maintain independent

Table 5.8
Comparison of Alternatives with Reported Timing Precision Needs on Networks

GPS Alternative	Communications		Electrical		Finance
	LTE-TDD 1–1.5 µs	Network Touters 4 µs	Phasor Measurement 4 µs	SPS/RAS 1 ms	High-Speed Trading 100 µs[a]
Legacy GPS					
GNSS DFMC					
eLoran					
Iridium-based					
Locata					
NextNav MBS					
NTP	Not close	Not close			Not close
PTP					
White Rabbit					
Holdover clock[b]					
LTE/4G cellular	Within factor of 5	Within factor of 5	Within factor of 5		
5G cellular					
WWVB	Not close	Not close	Not close	Not close	
DCF77	Not close	Not close	Not close	Not close	
DGNSS/other					
(Proprietary A)	Low TRL	Low TRL	Low TRL	Low TRL	Low TRL
(Proprietary B)	Low TRL	Low TRL	Low TRL	Low TRL	Low TRL
(Proprietary C)	Low TRL	Low TRL	Low TRL	Low TRL	Low TRL

▨ Meets precision   ▦ Precision within factor of 5   ■ Not close to required precision
▨ Low TRL—would require additional development

[a] Required for FINRA CAT NMS participants in the United States and for high-frequency trading in the EU.
[b] Meets accuracy requirement for a period of time depending on clock quality, but will eventually drift.

accurate timing have wired connections, so wired time delivery solutions should always be valid alternatives for consideration. Table 5.8 shows that while some legacy timing systems such as NTP and 4G cellular may be inadequate to meet modern timing requirements, all the new systems that have been proposed are capable of meeting the precision of all critical timing applications (assuming they are intended to deliver a timing signal at all). Precision performance in this case is not an important distinction between alternatives.

Some systems, including eLoran and Iridium-based solutions, have the advantage of wide area coverage so they could make a timing signal available even to scattered users in remote areas. On the other hand, White Rabbit technology could provide very precise synchronization between fiber-connected nodes of a network. NextNav MBS has very low-cost receivers and so would be a cost-effective way to synchronize timing among many sites in a region where it was available.

Table 5.9 examines reported positioning accuracy ranges for applications that occur primarily in urban areas or industrial or population centers. It is relatively likely that cell networks and other dense networks of transmitters will be available in these environments. In this case, there are meaningful differences in system performance because some PNT systems are able to meet all requirements while others cannot do

Table 5.9
Comparison of Alternatives to Reported Positioning Accuracy Needs in Urban Areas

GPS Alternative	Chemical	Commercial		Finance	Health
	Supply chain tracking 5 m	Construction <1 m	Location-based marketing <5 m	Asset tracking 15 m	Patient tracking <5 m
Legacy GPS					
GNSS DFMC					
eLoran[a]					
Iridium-based[a,b]					
Locata					
NextNav MBS					
WiFi/WLAN					
LTE/4G cellular					
5G cellular					
DGNSS/other					
Inertial nav.					
SLAM[c]	?	?	?	?	?
(Proprietary B)					

▦ Meets precision   ▦ Precision within factor of 5   ■ Not close to required precision
◨ Low TRL—would require additional development

[a] Assumes differential correction.

[b] Due to convergence time, Iridium-based position fixes would not be highly dynamic.

[c] Performance of SLAM is dependent on availability of landmarks and sensor data.

so or only have a marginal ability to meet some requirements. Of course, systems that only provide timing are not able to meet positioning requirements, so we do not list them as alternatives here.

As Table 5.9 shows, the Locata system and an additional proprietary system are capable of meeting all proposed positioning requirements. The MBS system, which would broadcast to cell phones, and navigation systems based directly on 5G or Wi-Fi signals, can meet most urban requirements except submeter requirements (e.g., for construction). Iridium-based and eLoran-based solutions have a somewhat marginal ability to meet urban positioning requirements, even though those systems would have the benefit of differential correction in urban areas.

Table 5.10 considers positioning applications that likely are or may be in remote areas. Although we continue to list navigation alternatives based on local transmitter networks, it may not be practical to build out those transmitter networks across undeveloped regions. As the table shows, the most demanding centimeter-level infrastructure monitoring requirements can only be met by Locata or an unnamed proprietary system, which would likely require an expensive point deployment of one of these systems. However, these monitoring requirements are not likely to be time critical, so the users would likely be able to deploy such a system or another surveying method if necessary.

Nevertheless, even setting aside those applications, as the table shows, there are no entirely satisfactory backups to GNSS available in remote areas. Iridium-based and eLoran-based solutions could cover wide areas and can meet some of the less demanding positioning requirements, but they would not be able to guide agricultural or drilling machinery with sufficient precision. Those solutions that do provide enough accuracy could not likely be deployed to cover large rural areas.

Finally, we consider how proposed PNT alternatives meet navigation needs in the transportation sector. Table 5.11 considers the requirements for performance-based navigation for aircraft. Aircraft en route do not require very high precision positioning, but only wide area systems such as eLoran or Iridium-based systems would likely provide coverage over those routes. Because an Iridium-based system requires time to accumulate satellite measurements before achieving a position fix, it would likely have poor performance in tracking the position of a fast-moving aircraft. Thus, of the systems considered here, eLoran is the only proposal (apart from GNSS) likely to provide navigation signals to aircraft throughout flight. Then again, eLoran does not provide a vertical approach capability and does not provide the level of accuracy generally required for precision landings. Local systems such as MBS or Locata could provide a high-precision navigation signal in the vicinity of an airport, though it would need to be approved for this use by the FAA and integrated into aircraft, which is usually a time-consuming process.

Table 5.12 compares the proposed PNT alternatives with reported needs for maritime and road users. eLoran and Iridium-based systems are the only proposals that

Table 5.10
Comparison of Alternatives with Reported Positioning Accuracy Needs in Remote Areas

GPS Alternative	Dams	Nuclear		Energy			
	Integrity monitoring 1 cm	Integrity monitoring 1 cm	Material tracking 1–5 m	Pipeline inspection <5 cm	Oil rig positioning <50 cm	Oil exploration 5 m	Oil supply 10 m
Legacy GPS							
GNSS DFMC							
eLoran with diff. corr.[a]							
Iridium-based[b] with diff. corr.[a]							
Locata[a]							
NextNav MBS[a]							
Wi-Fi/WLAN[a]							
LTE/4G cell[a]							
5G cell[a]							
DGNSS/other							
Inertial nav.							
SLAM[c]	?	?	?	?	?	?	?
(Proprietary B)							

GPS Alternative	Emergency Services		Agriculture			
	Vehicle tracking <1 m	Search and rescue 10 m	Planting 10 cm	Swath control 1 m	Food control (FDA) 5 m	Fertilizer application 30 m
Legacy GPS						
GNSS DFMC						
eLoran with diff. corr.[a]						
Iridium-based[b] with diff. corr.[a]						
Locata[a]						
NextNav MBS[a]						
Wi-Fi/WLAN[a]						
LTE/4G cell[a]						
5G cell[a]						
DGNSS/other						
Inertial nav.						
SLAM[c]	?	?	?	?	?	?
(Proprietary B)						

■ Meets precision   ▨ Precision within factor of 5   ■ Not close to required precision
◺ Low TRL—would require additional development
◿ Local signals—may not be available in remote areas

[a] If available, but it may not be available in remote areas.
[b] Due to convergence time, Iridium-based position fixes would not be highly dynamic.
[c] Performance of SLAM is dependent on availability of landmarks and sensor data.

Table 5.11
Comparison of Alternatives with Reported Aviation Accuracy Needs

GPS Alternative	Aviation							
	En Route	Terminal	NPA	APV-I	APV-II	CAT I	CAT II	CAT III
Horizontal	3700 m	750 m	220 m	16 m	16 m	16 m	7.5 m	3 m
Vertical	—	—	—	20 m	8 m	4-6 m	1 m	1 m
Legacy GPS[a]								
with SBAS/WAAS								
with GBAS[b]							c	c
GNSS DFMC								
with diff. corr.								
eLoran[a]								
with diff. corr.[a,b]								
Iridium[d]	X	X	X	X	X	X	X	X
with diff. corr.[a,b,d]	X	X	X	X	X	X	X	X
Locata[a,e]				e	e	e		
NextNav MBS[b,e]				e	e	e		
Wi-Fi/WLAN[a,b]								
LTE/4G cell[a,b]								
5G cell[a,b]								
DGNSS/other[a,b]								
Inertial nav.[f]								
SLAM[g]	?	?	?	?	?	?	?	?
(Proprietary B)								

■ Meets precision   ▨ Precision within factor of 5   ■ Not close to required precision
◫ Low TRL—would require additional development
▨ Inadequate signal range   ☒ Poor performance with moving receiver

NOTES: SBAS = satellite-based augmentation system; WAAS = wide area augmentation system; GBAS = ground-based augmentation system.

[a] Does not provide a vertical position with adequate integrity for aviation use.

[b] Where available.

[c] Standards for a GBAS CAT II/III capability are complete, but such a ground system has not yet completed certification although extensive testing has been accomplished.

[d] Due to convergence time, Iridium-based position fixes would likely fail to track moving aircraft.

[e] Locata and NextNav could provide adequate 95 percent accuracy and so might be able to support landings to CAT I. However, more study and testing would be required to demonstrate that either of these systems could meet the required aviation integrity assured accuracy bound, provided with high availability and without continuity disruptions.

[f] Inertial navigation can support en-route and terminal operations, but inertial drift limits the duration of utility for operations. All aviation-grade navigation inertial reference units are assumed to support a drift rate of 2.0 nonmaskable interrupts (nmi) per hour (or better).

[g] SLAM is an emerging/experimental technology. Performance could be good in principle, but has not been demonstrated.

Table 5.12
Comparison of Alternatives to Navigation Needs in Maritime and Road Sectors

GPS Alternative	Maritime		Road Vehicle to Infrastructure		
	Open Water 10 m	Port 1 m	Road 5 m	Lane 1.1 m	Where in Lane 0.7 m
Legacy GPS					
GNSS DFMC					
eLoran					
with diff. corr.[a]					
Iridium[b]	X	X	X	X	X
with diff. corr.[a,b]	X	X	X	X	X
Locata[a]					
NextNav MBS[a]					
Wi-Fi/WLAN[a]					
LTE/4G cell[a]					
5G cell[a]					
DGNSS/other[a]					
Inertial nav.					
SLAM[c]	?	?			
(Proprietary B)					

■ Meets precision   ▨ Precision within factor of 5   ■ Not close to required precision
◹ Low TRL—would require additional development
◺ Inadequate signal range   ⊠ Poor performance with moving receiver

[a] If available, but may not be available in remote areas.

[b] Due to convergence time, Iridium-based position fixes would not be highly dynamic.

[c] Performance of SLAM is dependent on availability of landmarks and sensor data.

would provide coverage in the open ocean, albeit with less position accuracy than desired. The Iridium-based system would provide static position fixes and performance would be degraded for a fast-moving ship. Neither of those systems can provide the accuracy for precision approaches into port, but Locata or NextNav MBS could. 5G cellular signals might also be sufficient for port navigation because a vertical position is not required.

For road users, eLoran and Iridium-based systems can only marginally meet the required precision to locate a vehicle on a road. In developed areas, NextNav MBS or 5G cellular service could provide turn-by-turn navigation. Only Locata, an unnamed proprietary system, or SLAM-based solutions could provide high enough accuracy to compare a vehicle's position with the center of its lane. SLAM-type solutions using cameras and/or lidar are employed by self-driving cars under development.

Summary

One important alternative provides a nearly seamless backup for GPS—the other GNSS constellations that use similar signals in the same radio band. User equipment is rapidly integrating all these. There has long been a desire in the United States (legally required in some cases) not to depend on the Russian GLONASS or Chinese BeiDou systems, but the current deployment of Galileo by the EU would offer a reliable and trustable backup. However, when something like a large solar storm disrupts the ionosphere, and so distorts GPS signals, these signals would also be distorted.

Other terrestrial backup and complementary positioning and navigation systems fall into two categories—systems that could affordably cover the entire nation but that lack the precision for many applications and so are of more limited value and systems matching or exceeding GPS accuracy but that can only be affordably deployed in small areas, such as ports, or in limited urban areas.

In contrast, there are a wide variety of alternative terrestrial timing systems that can support fixed locations, either broadcasting timing through radio or over fiber optics or else simply providing local clocks. Additionally, the terrestrial positioning and navigation systems also offer accurate timing over their areas.

Conduct Cost-Benefit Assessment of Alternative Sources and Technologies to Increase National PNT Resilience and Robustness

.

The last several chapters have developed most of what we need—the inputs—to consider the cost-benefit ratio for various alternatives, backups, and complements that could be added to the national PNT ecosystem. What remains is to consider how the different options and their ability to address the needs of low-to-high fidelity PNT across different sectors could help those sector avoid losses from the different GPS threats and hazards. Then, determining whether different options could be justified on an economic basis entails considering the number and likelihood of such disruptions in GPS that would be needed to make any particular investment a good societal choice.

To frame this consideration of costs and benefits, it is worth beginning with the conclusion from our economic analysis in Chapter Four: *When estimates of the cost of GPS disruption or loss include realistic adaptation options and existing complementary technologies (backups and workarounds), the estimates are quite modest and are often much lower than what we have seen in other previous analyses.* Given this conclusion, the bar for government investment in supplements to the existing PNT ecosystem would be quite high. The conclusion from our economic analysis differs from other previous analyses that have considered the costs of PNT more in isolation and have focused on low-probability scenarios that would produce large-scale and enduring outages and that have therefore produced much larger losses than we have estimated.[1]

The challenge in showing positive cost-benefit for new systems added to the ecosystem to increase PNT robustness and resilience is further complicated by the fact that there are programs working toward significant national PNT augmentations driven by other requirements and policy initiatives. For example, as discussed in Chapter Five, technologies that are under consideration for integration into FirstNet to meet E911 and first responder location requirements will, if implemented, also be full terrestrial PNT backup systems in the areas where they are put in place. The system under consideration provides both positioning and navigation at a level com-

---

[1] In Chapter Seven, we will consider when a belief in these much larger estimates would change our conclusions.

parable or better than GPS, involves distributed clocks that can provide timing, and provides a signal that is already compatible with some consumer-level user devices. Although the focus of the program is on urban areas where the need for E911 and responder location within the built environment is greatest, such areas are also concentrated nodes of both population and economic activity. As a result, justifying government funding and implementation of additional systems beyond what is being done under FirstNet—to cover, for example, suburban and more rural areas—would have only the residual economic benefit not covered by the FirstNet initiative to balance against their costs.

This chapter examines the final part of our analysis: ***conduct cost-benefit assessment of alternative sources and technologies to increase national PNT resilience and robustness.*** In doing that assessment, we sought to answer the questions shown in Table 1.1: how do the costs of available backups compare with the benefits those backups would support? In particular, which, if any, have costs less than or comparable to their benefits? What are the opportunities for public-private partnerships (PPPs) in supplying the promising backups? In this discussion, we keep the question of "who pays" for any required investment somewhat open, considering both the drivers of overall system costs and what factors could cause those costs to be borne by government.[2] ***Our bottom line finding is that the modest damages from GPS outages do not alone justify the cost of building another large-scale backup system. That said, the federal government is already deploying a system through FirstNet to provide a PNT signal indoors and vertically. The real justification for that system is to provide floor-level information on the location of first responders within high-rise buildings; however, it inherently will also serve as a backup during rarer space storms and jamming events.***

In the remainder of this chapter, we take on this issue of cost-benefit balance, first presenting a summary set of daily loss estimates based on our economic analyses, then examining the relative costs associated with different systems, and finally looking at the balance between potential loss avoidance and system costs given the application of the options across infrastructures and sectors.

Considering the Costs of GPS Disruptions

In Chapter Four, we discussed three approaches for making estimates of the potential losses from a disruption to GPS. In considering the threats and hazards affecting GPS, most scenarios—both natural and adversarial—focus on outages of hours or days, where even solar storms are viewed as more likely to result in transient disruption of the

---

[2] The question of PPPs is briefly discussed here, but it is discussed in more detail in Chapter Seven and in Appendix D.

system (potentially coupled with reduced life spans of on-orbit satellites) than immediate destruction of major parts of the constellation.[3] However, the nature of scenarios differ in whether they would affect the country (e.g., space weather) or a localized area (e.g., most operationally practical jamming or spoofing scenarios).

For considering countrywide effects like space weather, the results of the nationwide analyses provides one set of estimates of the costs associated with disruption, with the results presented in Chapter Four, whether anchored in literature values adjusted as previously described, aggregating our regionally based estimates, or drawing on the O'Connor et al. (2019) analysis.

Table 6.1 reproduces Table 4.1 and adds the total per day value ranges from that analysis that we will use here, inflated to 2018 dollars. In each row, we have highlighted in color the cell we use in the final column for our further estimates of costs and benefits. For several of the rows, we have adopted the result from O'Connor et al. (2019) as the best nationwide estimate. We used the estimate described earlier in this work for emergency services, as that was the only explicit one we found. We show a range for port operations to explicitly include the highly uncertain potential for large costs. Several categories—Finance, People Tracking, Aviation, and Railway—are generally small, whichever range we used. Finally, we do not consider further in this chapter the high estimate of damages in the O'Connor et al. (2019) analysis from the loss of timing for wireless telecommunications—the cellular system. As discussed earlier, this is so inconsistent with our findings from the operators of the systems that we do not feel it is suitable for informing decisionmaking. Nonetheless, we return to the general issue of a belief in very high damages in Chapter Seven.

Across the economy, the total estimate we will carry forward thus covers a loss range from just over $785 million up to $1,318 million per day.

Not all of these damages will be relevant to our cost/benefit calculation. First, none of the systems under consideration can mitigate all of these damages. The most striking example of that is precision agriculture. While the range in damage estimates from agriculture is wide and discussed in Chapter Four at some length, none of the systems under consideration can affordably cover such wide areas as agricultural belts while providing the submeter precision that is required. Consequently, whichever of these estimates of damages in the agricultural sector due to a CME or similar nationwide disturbance turns out to be correct does not affect our cost/benefit estimate. For this reason, whether we take the worst-case or average-case number for outage costs to the agriculture sector makes no difference to the overall argument.

Another potentially large damage estimate comes from disrupted container port activities, again discussed in more depth in Chapter Four. Since the uses in port opera-

---

[3]   As discussed previously, the main exceptions to this are scenarios involving nuclear war in space (that would affect all satellite-based GNSS systems) or some type of destructive cyber intrusion that could have focused effect on the GPS system alone.

Table 6.1
Estimates of Daily Cost of Losing GPS Services Nationwide (millions of then-year dollars)

Sector	O'Connor (2017$)	Pham (2011$)		Leveson (2013$)	Extrapolated Range from Leveson and Pham (2018$)			HSOAC (2018$)	Likely Range (2018$)
					Low	Medium	High		
Consumer location-based services	95	N/E		36	44	128	254	N/E	97
Commercial road transport	138	28	28	33	20	61	123	N/E	141
Emergency services	N/E[a]	N/E		N/E	N/E	N/E	N/E	11–72	11–72
Agriculture	503[b]/41[c]	553[b]/54[c]	553[b]/54[c]	381[b]/38[c]	220[b]/22[c]	340[b]/34[c]	500[b]/49[c]	N/E	514[b]/42[c]
Construction	N/E	25[d]	10[e]	14	8	24	43	345	24
Surveying[f]	11[f]		15[e,g]	32[g]	14[g]	15[g]	17[g]	7[f]/11[g]	7[f]
Aviation	N/E	77[d,e]	7[e]	0.4	4[e]	6[e]	9[e]	N/E	6
Maritime	123[h]		61[e]	0.5	36[e]	58[e]	85[e]	N/E	126
Railway	N/E		0.1[e]	0.1	0.1[e]	0.1[e]	0.2[e]	N/E	0.1
Timing (telecom)	40–456[i]							0	0
Timing (electric grid)	9		9[e]	0.1[j]	6[e]	9[e]	13[e]	N/E	9
Timing (finance)	0							9	9
People tracking	N/E		0.6[e]	N/E	0.3[e]	0.6[e]	0.8[e]	N/E	0.7
Port operations	224[k]	N/E		N/E	N/E	N/E	N/E	0–880	229
Mining	32	N/E		N/E	N/E	N/E	N/E	N/E	32
Oil and gas	51	N/E		N/E	N/E	N/E	N/E	N/E	51
TOTAL									785–1,318

NOTES: N/E = not estimated. Note also that while these numbers are often presented with two or three significant figures, that is only to ease any numerical comparison across this document; the actual precision of all these estimates is much less.

[a] O'Connor et al. (2019) decided not to quantify the benefit of lives saved by emergency responders, but did assess consumers' willingness to pay for more accurate emergency call location as a part of location-based services.

[b] Daily cost for an outage occurring at a critical period of agricultural activity (planting season).

[c] An approximate expected daily cost for an outage occurring at a random time of year.

[d] Pham (2011) combined these sectors when estimating GPS benefits.

[e] This estimate rests on an assumption that GPS benefits are roughly proportional to GPS spending across sectors, based on Pham's (2011) data from 2005–2010.

[f] Excludes surveying for the mining and oil and gas sectors.

[g] Includes surveying for the mining and oil and gas sectors.

[h] Includes recreational boating (the dominant component), commercial fishing, navigation in seaways, and towing. Does not include cargo operations in ports, which are listed separately.

[i] In the O'Connor et al. (2019) model, costs would start at $40 million on the first day and then escalate, to $456 million per day after 30 days.

[j] Leveson (2013) estimated very small timing benefits from GPS, on the assumption that an alternative global time system could be built, but these numbers are not meaningful for a sudden GPS outage with no global alternative in place.

[k] Average daily impact from a 30-day outage. In the O'Connor et al. (2019) model, daily impact would initially be much less, but would grow larger with the duration of the outage.

tions also require submeter accuracy, these damages can only be mitigated by one local system, Locata, among the systems we analyzed. As noted in the O'Connor et al. (2019), 90 percent of the damages comes from disruptions in just four ports—Long Beach, Los Angeles, Norfolk, and New York. Consequently, the most practical alternative if the high estimate concerns a decisionmaker would be the installation and integration of some commercially available, GPS-independent, automation system at just these four ports, a comparatively low cost. Locata is the one system included this report that does already support operations at some international ports, but other alternatives exist that are specifically designed for automating container port operations (Alho et al., 2015). That does not address the question whether this is an appropriate investment of federal funds or whether this should instead be left to the port operators, as such systems are intended to provide efficiency benefits to port operations.

All the above concerns nationwide outages of GPS. However, as described in Chapter Three, countrywide disruptive scenarios represent only a portion of the threat and hazard space relevant to considering GPS risk. Smaller-scale incidents, whether driven by what we labeled "negligent threats" or intentional adversarial jamming efforts, can be staged locally. Our microlevel estimates of costs associated with such incidents focused on specific geographic locales and were selected in an effort to pick targets in particular industries that where an attacker might expect to maximize losses (e.g., we selected areas with concentrated economic activity in the sectors of interest) (see Table 6.2).

The estimates of the effect of smaller-scale incidents can in some cases begin at zero, reflecting the challenge in timing and targeting jamming operations to affect sensitive activities in some sectors. Our expectation is that the costs of localized events are more likely to fall near the low end of our range, suggesting locally disruptive events producing effects in the very low tens of millions in cost in most cases and frequently much lower levels of damage.

Table 6.2
Estimated Daily Cost of Losing GPS Services in Targeted Regional Jamming Scenarios (millions of 2018 dollars)

Targeted Sector	Scenario Location	Estimated Daily Cost (millions of 2018 $)
Consumer location-based services	Chicago	3.2
Agriculture (wheat)	Kansas	0.4–1.5
Construction	Houston	5
Surveying	Atlanta	0.05
Finance	Los Angeles	1
All sectors equally targeted	Los Angeles–Long Beach	5–13

Considering the Relative Costs of PNT Options

Because cost information associated with many technologies or options that could be added to the national PNT ecosystem are proprietary to the firms providing those technologies, we are unable to publicly report detailed calculations of costs for implementation of systems across different portions of the country.[4] However, because of the range of options that could be acquired or built on to increase the robustness and resilience of national PNT, we make some distinctions about relative cost only at a qualitative level. Significant differences in cost among options exist based on whether the technologies are relevant at the individual user level or are systems intended to provide PNT broadly. Also, there are technologies built and maintained for commercial reasons wholly separate from PNT provision; they require limited to no government support or even involvement but play their roles in the GPS ecosystem.

Tables 6.3 and 6.4 examine the cost drivers for each of the complementary technologies for positioning and timing. Cost drivers are broken down into the following categories.

- **New space infrastructure needed:** Because new space systems are costly, the requirement for construction of new space infrastructure would be a significant increase in system cost. To the extent that infrastructure already exists (e.g., the existing GPS constellation), we do not count it for this forward-looking characterization.
- **New terrestrial infrastructure needed in covered areas:** For systems that depend on ground-based transmitters or beacons, the cost of putting such infrastructures in place can be significant. This includes systems like eLoran, whose infrastructure is designed to cover larger areas, and also systems designed for smaller footprints based on smaller beacons. To the extent that ground-based infrastructure already exists (e.g., existing cell tower network built and maintained privately to support mobile telephony, not PNT), we do not count it.
- **Individual user costs to utilize for PNT:** If any complementary technology is to make a measurable contribution to PNT robustness and resilience, it must be adopted and used. To the extent that individual users must pay a cost to do so—to buy a new receiver device, to pay for access to the signal, or acquire complementary technology like nulling antennas—adoption will be slower. On a national basis, even costs of tens to hundreds of dollars per user could add up to substantial costs over relevant user bases.

These characteristics together shape the overall national costs of the technology option as a PNT complement or backup, regardless of whether those costs are paid by

---

[4]   For detailed cost calculations for a subset of the potential technology systems, see the proprietary annex to this report.

Table 6.3
Cost Drivers: Complementary Positioning Technologies

Complementary Technology	New Space Infrastructure Needed?	New Terrestrial Infrastructure Needed in Covered Areas?	Individual User Costs to Utilize for PNT (including new devices or costs in addition to use of existing devices)	Expected USG Expenditure to Build/ Maintain as Core Part of National PNT Ecosystem
Multi-GNSS	No—Existing systems	No	No—Existing compatibility	None—Maintained by other nations
Wi-Fi/WLAN	No	No	No—Existing compatibility	None—Individual and commercial systems
LTE/4G cellular	No	No	No—Existing compatibility	None—Commercial systems
5G cellular	No	Yes, but Commercially Driven	No—Existing compatibility for some uses Yes for specialized users	None—Commercial systems
Inertial navigation	No	No	Yes—Individual acquisition	None—Low/medium, where cost would only involve individual subsidy to adopt[a]
Iridium-based	No—Existing System	No	Yes—Specialized receivers + access costs	Low[a]
Locata	No	Yes	Yes—Specialized receivers/no compatibility for many devices	High, if pursued as broad addition to national PNT ecosystem[a]
NextNav MBS	No	Yes—Beacons installed in covered areas	No—Compatibility being integrated into devices	High, if pursued as broad addition to national PNT ecosystem [a,b]
eLoran	No	Yes—Tower construction	Yes—Specialized receivers/no compatibility for many devices	Medium
SLAM	No	No[c]	Yes—Individual sensor platforms	None—Low[c]
Nulling antennas	No	No	Yes—Individual choice to adopt	None—Low/medium, where cost would only involve individual subsidy to adopt[a]

Table 6.3—Continued

Complementary Technology	New Space Infrastructure Needed?	New Terrestrial Infrastructure Needed in Covered Areas?	Individual User Costs to Utilize for PNT (including new devices or costs in addition to use of existing devices)	Expected USG Expenditure to Build/ Maintain as Core Part of National PNT Ecosystem
Direction finders	No	No	Yes–Individual choice to adopt	None–Low/medium, where cost would only involve individual subsidy to adopt[a]
Integrity monitoring/ jamming detection	No	Yes–Detector installation	N/A[d]	None–Low[e]

[a] Because these systems currently exist on the market, individual users can choose to utilize them and pay full costs for acquisition and operations/usage. Beyond that existing availability (where firms with market incentive for robust PNT can choose to adopt), a national decision to pursue these technologies as larger elements of the national ecosystem would require investment to broaden implementation and could involve subsidies to defray costs of adoption at the user level.

[b] Federal government may invest in deployment of NextNav MBS as a component of FirstNet to satisfy E911 and emergency responder requirements. If this action is taken, this technology will become available in portions of the country for reasons other than economically driven factors. Such a deployment would cover a significant portion of the U.S. population and centers of economic activity.

[c] SLAM technologies rely on navigation cloud databases which must be built and maintained which could be funded by government and provided as a public service to incentivize adoption could also be built by private technology developers driven by the market for autonomy.

[d] Detection technologies do not have the same end-user populations as PNT-providing technologies. To enable rapid detection and interdiction of jamming, detection tools could be provided to local law enforcement organizations. The numbers of end users for such technologies would be exponentially smaller than that for PNT-utilizing devices.

[e] Because jamming is criminal violation punishable by a fine, detection infrastructure could be revenue producing or installed/maintained commercially as is the case for some traffic enforcement technologies.

individual users, the private sector or government. However, for some systems—particularly those where some or all of their functionality is already available—some categories of costs can be definitively assigned to specific economic actors. Given that the focus here is U.S. government actions that could strengthen national PNT robustness and resilience, we bring together the cost components into approximate "potential U.S. government cost," based on what could be involved for the technology to make a significant contribution to national PNT robustness and resilience (i.e., beyond incidental or localized use by individual firms or sectors). These costs could be direct government costs (e.g., appropriations to build eLoran or implement local positioning systems in multiple locations) or indirect (e.g., subsidies to users to promote use of nationally desirable PNT options to strengthen their role in the ecosystem).

Table 6.4
Cost Drivers: Complementary Timing Technologies

Complementary Technology	New Space Infrastructure Needed?	New Terrestrial Infrastructure Needed in Covered Areas?	Individual User Costs to Utilize for PNT (including new devices or costs in addition to use of existing devices)	Expected USG Expenditure to Build/ Maintain as Core Part of National PNT Ecosystem
Multi-GNSS	No–Existing systems	No	No–Existing compatibility	None–Maintained by other nations
NTP	No	No	No–Part of existing technologies	None
PTP	No	No	Yes–Individual implementation	None–Medium, depending on role promoting use
Holdover clock	No	No	Yes–Individual Clock purchase	None–Medium, if pursued as broad addition to national PNT ecosystem[a]
LTE/4G cellular	No	No	No–Existing compatibility	None–Commercial systems
5G cellular	No	Yes, but commercially driven	No–Existing compatibility	None–Commercial systems
Iridium-based	No–Existing system	No	Yes–Specialized receivers + access costs	Low[a]
Locata	No	Yes	Yes–Specialized receivers/no compatibility for many devices	High, if pursued as broad addition to national PNT ecosystem[a]
White Rabbit	No	No	Yes–Specialized devices commercially available	None–Medium, if pursued as broad addition to national PNT ecosystem[a]
NextNav MBS	No	Yes–Beacons installed in covered areas	No–Compatibility being integrated into devices	High, if pursued as broad addition to national PNT ecosystem[a,b]
eLoran	No	Yes–Tower construction	Yes–Specialized receivers/no compatibility for many devices	Medium
Nulling antennas	No	No	Yes–Individual choice to adopt	None–Low/medium, where cost would only involve individual subsidy to adopt[a]

Table 6.4—Continued

Complementary Technology	New Space Infrastructure Needed?	New Terrestrial Infrastructure Needed in Covered Areas?	Individual User Costs to Utilize for PNT (including new devices or costs in addition to use of existing devices)	Expected USG Expenditure to Build/ Maintain as Core Part of National PNT Ecosystem
Integrity monitoring/ jamming detection	No	Yes–Detector installation	N/A[d]	None–Low[e]

[a] Because these systems currently exist on the market, individual users can choose to utilize them and pay full costs for acquisition and operations/usage. Beyond that existing availability (where firms with market incentive for robust PNT can choose to adopt), a national decision to pursue these technologies as larger elements of the national ecosystem would require investment to broaden implementation and could involve subsidies to defray costs of adoption at the user level.

[b] Federal government may invest in deployment of NextNav MBS as a component of FirstNet to satisfy E911 and emergency responder requirements. If this action is taken, this technology will become available in portions of the country for reasons other than economically driven factors. Such a deployment would cover a significant portion of the U.S. population and centers of economic activity.

[d] Detection technologies do not have the same end-user populations as PNT-providing technologies. To enable rapid detection and interdiction of jamming, detection tools could be provided to local law enforcement organizations. The numbers of end users for such technologies would be exponentially smaller than that for PNT utilizing devices.

[e] Because jamming is criminal violation punishable by a fine, detection infrastructure could be revenue producing or installed/maintained commercially as is the case for some traffic enforcement technologies.

In considering overall costs, installation of new infrastructure—whether space-based or terrestrial—would be major cost drivers. None of the systems currently available involve entirely new space-based infrastructure, but several involve terrestrial infrastructure. Constructing that infrastructure would vary from systems relying on fewer but more significant sites (e.g., eLoran) versus those that use beacon systems to cover geographic areas (e.g., Locata, NextNav MBS). Although specific values for individual systems are proprietary,[5] coverage of any significant portion of the country (e.g., hundreds of metropolitan areas, wide area coverage for eLoran) would involve investments in the hundreds of millions to low billions of dollars, depending on the system. Operations and maintenance costs for systems fall over a similar range. For technical reasons described in more detail in the appendix, timing-only systems are much less costly than positioning/navigation systems, with all options falling near the bottom of the ranges suggested here.

Costs associated with building and maintaining infrastructure are centralized (e.g., if government procured any complementary system, the costs of putting infrastructure in place would be easy to see). Costs associated with the equipment needed

---

[5]   See the proprietary Appendix G of this report.

to *use* the PNT data provided by the systems (or the costs associated with resilience investments like nulling antennas) would be more diffuse, because they would largely be paid by individual users. Although the "per unit" cost of user equipment might not be large, when aggregated across the large number of users for some PNT applications, the costs would add up to nationally significant values. For some technologies, receiver costs are currently substantial: As custom and specialized devices, their costs can reach into the thousands of dollars. Although such costs could go down if demand and miniaturization resulted in economies of scale, even low costs per receiver add up rapidly across a potential user base numbered in the millions as for consumers, the largest single set of users.

Considering the Potential Benefits from Different PNT Options

As the last two chapters have made clear, no single system is a perfect backup for GPS. Table 6.5 shows which proposed PNT solutions are the strongest candidates to meet the needs of different types of users, based on their accuracy, their practical coverage areas, and their ability to be made available to users at low cost.

Example Cases of Potential Losses Avoided from Threats and Hazards Compared with Costs

In this section, we drill down and look specifically at example cases of the benefits (avoided losses) versus the costs of such complementary systems for specific threats and hazards.

Nationwide Outage from Solar Storm

We now can total the costs and potential benefits, starting with a nationwide outage from a solar storm when the CME disturbs the ionosphere for a day, disrupting all GNSS signals. In particular, this means that Galileo and GLONASS would also not be usable. We use the various economic cost estimates from Table 6.1.

From the tables above, the loss estimates, and our cost estimates, we can construct Table 6.6 for nationwide losses avoided for three days of such a storm. In the table, a check mark indicates the potential of this technology to backup GPS in that sector. For the estimate, we are crediting these technologies with the full value of sectors that they can even come close to supporting (i.e., "yellow" in the tables from Chapter Five and not ruled out for other reasons such as being impractical to deploy across wide areas). The exception is the question mark under aviation; there is technical potential, but since there is no existing or planned regulatory approval within the FAA for the use of these signals, they will not, in practice, be available for many years after deployment.

Table 6.5
Strongest Candidate PNT Systems for Different Critical User Segments

Use Case	A. Precision Timing on Networks	B. Moderate- Accuracy Positioning of People and Road Vehicles	C. High-Accuracy Positioning of Equipment	D. High- Reliability Positioning in Aircraft Landing	E. Low-Accuracy Positioning of Aircraft and Ships En Route
Critical sectors	Telecom, electricity, financial (see Table 5.1)	Road transport, government, commercial, many other sectors (see Table 5.2)	Agriculture, construction, dams, energy, maritime, nuclear, water (see Table 5.3)	Aviation (see Table 5.4)	Aviation, maritime (see Table 5.5)
Value of indoor coverage	**High**	Moderate	Moderate	Low	N/A
Importance of low-cost receivers for user adoption	Moderate	**High**	Low	Low	Low
Need for horizontal position accuracy	N/A	Moderate	**High**	Moderate	Low
Need for vertical position accuracy	N/A	Moderate	Moderate	**High**	Low
Importance of coverage far from populated areas	Low	Low	Moderate	Moderate	**High**
Strongest candidate PNT systems	Nationally: eLoran Satelles STL White Rabbit Regionally: NextNav MBS PTP	NextNav MBS	Locata	Existing instrument landing systems Possibly: Locata NextNav MBS	eLoran Satelles STL

We thus do not include that financially small contribution in the total. The right-hand column totals the avoided damages over three days for each alternative.

Several sectors have no check marks; that is to say that none of these systems can replace GPS in that sector during this contingency. Consequently, those losses do not affect the decision to deploy or not deploy these systems.

Table 6.6
Preventable Losses of Potential Deployments for an Extreme (3 Day) Solar Storm (millions of 2018 dollars)

Sector	Consumer LBS	Road Freight	Emergency Services	Agriculture	Construction	Surveying	Aviation	Maritime	Railway	Telecom	Electric Grid	Finance	People Tracking	Port Operations	Mining	Oil and Gas	3-Day Total
Daily disruption costs	97	141	11–72	42–514	24	7	6	126	0.1	0	9	9	0.7	229	32	51	
Multi-GNSS																	0
eLoran (timing only)										✓	✓	✓					54
eLoran (PNT)								✓		✓	✓	✓					432[a]
Iridium-based								✓		✓	✓	✓					432[a]
NextNav MBS		✓	✓				?			✓	✓	✓	✓				804–987
Locata						?								✓	✓		783[c]
5G cellular		✓	✓							N/A	✓	✓	✓	✓	✓		804–987[b]
PTP										✓	✓	✓					54
White Rabbit										✓	✓	✓					54
Holdover clock										✓	✓	✓					54

NOTES: While these numbers are often presented with up to three significant figures, that is only to ease any numerical comparison across this document; the actual precision of all these estimates is much less. LBS = location-based services.

a The savings attributed to both eLoran and the Iridium-based solutions for positioning depend entirely on the adoption of its receivers by recreational boaters, historically a cost-sensitive market.

b The potential of 5G as an alternative for timing and/or positioning depend on its own independence from GNSS, which is not yet definite.

c The savings attributed to Locata is dominated by potential losses in just four ports.

Note that this does not include any estimate of the likelihood of this event. As noted earlier, the frequency of a solar storm of the magnitude of the Carrington event hitting the earth is hard to estimate with precision but appears to be roughly once a century. That makes the damages that might be expected in any decade to be (also roughly) about one-tenth of the amounts above.

We now need to compare these damages with the costs for a range of alternative technologies. As we discussed previously, these costs range from the hundreds of millions into the billions of dollars for high cost systems over a decade to less for individually purchased resilience technologies.

What is clear is that taking that frequency into account, the scale of the expected value of the avoided damages remains well below the cost for dedicated PNT alternative systems. In part, this is a simple consequence of the inability of alternatives to practically support some uses, notably agriculture. Additionally, some potential damages are amenable to being mitigated by comparatively small investments—most notably, the ability to mitigate damages at container ports by deploying high-accuracy backups in only four ports. Finally, this conclusion is further buttressed by several existing and likely future components of the PNT ecosystem for which little or no new government investment is required; deployments of the FirstNet location system within urban areas and the 5G network will provide backups in some urban areas, thereby further weakening the case for any significant federal investment in additional alternatives.

Local Outages from Jamming or Spoofing

Jamming and other local attacks are by nature significantly less damaging than the nationwide loss of PNT, as summarized in Table 6.2. Across industries, summing the lower ends of the estimated range produces overall estimates in the low single to tens of millions, with an upper-end tail reaching into the hundreds of millions only if very significant transportation effects on the order of large-scale natural disasters occur with GPS disruption, effects we found no evidence to support in the existing cases of GPS disruption or our regional analyses. Such low estimates of costs for individual incidents would seem to argue that if any investment is made in response, that investment should be at the low end of the spectrum in resilience technologies; such technologies include direction finding, integrity monitoring, and other tools to make it possible to rapidly detect jamming, identify the perpetrators, and halt their activities. To the extent that the proliferation of such capabilities makes jamming nonviable as a consequential operation (and increase the cost associated with negligent disruption), the deterrent effect could reduce risk disproportionate to the costs.

Our analysis of jamming does not include a case of jamming across the entire country on a sustained basis. That is beyond the capabilities of all conceivable threat actors, even those supported by hostile nation-states. However, commercial systems can approximate that case, in particular any deployment of strong transmitters in or near GPS bands for other purposes.[6] Such deployments could have much more serious effects than those explored here. Certainly, the direct disruption and loss would be greater and sustained indefinitely. Additionally, any specific circumstances in which the loss of GPS could have outsized costs, such as by triggering rare, difficult to predict effects, would in fact actually occur and cause losses because of the nationwide coverage and continuing transmission of such systems. The analysis in this report does not apply to such emitters, but it seems that guarding against such threats through actions to safeguard the spectrum used by GNSS should remain a priority to avoid running such a natural experiment on the U.S. economy.

Summary

We focused on near-term technologies, ready for deployment, with little or no need for further development. In general, we find that no threat or hazard would cause likely damages even close to the cost of deploying a backup or alternative PNT system. This held true for both positioning and navigation and for timing, together or separately, nationally or regionally, and considering all uses of GPS across sectors, from transportation to E911.

To argue strongly for deploying any of these systems purely as a backup to GPS—independent of whether such systems are funded by the government or private industry—requires believing one (or both) of two alternatives. In the first, one must believe that the threats to GPS are much worse than these. That is, the outages would be widespread, frequent, and last for many days. Additionally, one generally must believe the damages would be close to the upper estimates that we list, not the lower. In the second alternative, one would need to believe in the much higher estimates of damages, such as the high estimates for telecommunication disruption in the O'Connor et al. (2019) analysis.

Important assumptions are being made here that are key to this finding, but we believe they are reasonable ones. First, we assume that no industry will discard their existing backups and simply rely on GPS. Given the occasional problems with GPS (nuisance jammers, a satellite with an incorrectly loaded ephemeris, etc.), not relying on GPS alone would seem to be well motivated.

Second, we similarly assume that systems using GPS critically will be designed to monitor the integrity of GPS information. Such integrity monitoring would obviate spoofing attacks, reducing their effectiveness to near zero. More importantly for this analysis though, the ability of a user to do integrity monitoring would not be changed by providing another alternative or backup signal. This is because in all the cases that we considered, there already are abundant available checks, whether from other signals

---

[6] See the DoT (2018b) for a technical analysis of such interference.

such as from other GNSS and cellular towers and/or from internal subsystems, such as other sensors such as radar, an IMU, or simply a known, fixed location as for the financial or electric grid uses. Adding another signal does not change that—if a system does not use the existing systems to monitor integrity, it is unlikely to use yet another.

Finally, and independent of any need to backup GPS, the federal government is already deploying a partial backup. AT&T FirstNet—a PPP that supplies robust cellular communications to first responders—will deploy a backup that independently provides accurate positioning (including height) indoors in most built-up urban areas. AT&T FirstNet is using a signal compatible with cell phones and will also provide accurate timing. Consequently, this signal will cover a large fraction of the U.S. populace and a great deal of economic activity, which depending on the scale of the FirstNet deployment of their accurate positioning further reduces any benefit from other deployments.

Of course, some of the systems may well be deployed independent of any concern about GPS outages. For example, Locata is being deployed in ports and mines for their internal use, and as noted, MBS or some similar system will be deployed to support z-axis determination in some areas for first responders and for those areas will enable location-based services. SLAM will likely be used and matured by developing autonomous vehicle and drone applications. We discuss this in the final chapter.

CHAPTER SEVEN

Summary and Weighing Alternatives and Potential Federal Initiatives

In this final chapter, we provide a brief overview of the findings from Chapters Three through Six that cover the four parts of our analysis. Then, we discuss different ways to interpret those findings based on individual worldviews. Finally, we examine some potential federal initiatives.

Summary

In Chapter Six, we found that the generally modest damages that we expect from GPS disruptions do not alone justify the cost of a backup. Still, different decisionmakers often hold quite different views of threats and the importance of different domestic infrastructures. In this chapter, we try to accommodate such divergent views by asking, "What must one believe in order to change that conclusion?"

For those decisionmakers most concerned with local jamming, whether by terrorists or state-supported groups, the limited area affected by any such jamming yields a perhaps predictable implication: One would need to believe that the number of such events was likely to consistently be at least several score a year—far beyond the utter absence of such attacks to date—in order to justify any new system.

For those more concerned with larger-scale, national outages or disruptions, requiring a nationwide positioning (and thus navigation) backup capability could be driven by one of two sets of beliefs. First, decisionmakers must believe that the damages would near the high end of our estimates *and* that such events would occur several times per decade or last much longer than we expect the effects of a solar storm to last. Alternatively, they would need to believe that some very large damage estimate is in fact correct—for example, that container ports would be rendered inoperable. As we have seen, such beliefs would lead to supporting the deployment of systems providing intermediate or higher accuracy in urban areas to prevent most of the damage, or to support a sector where they believe large damages would occur, such as container ports.

For those more concerned with a national-level outage and *timing*, decisionmakers must believe that the losses from a disruption in timing would be much larger than our estimates. This might entail believing that existing backups would be abandoned, producing larger damages in such sectors as finance, or believing that damages would in fact be much higher—for example, the very high estimates in O'Connor et al. (2019) for the wireless communication sector. These cases could motivate the deployment of any of a wide variety of relatively inexpensive alternatives, from satellite-based systems (notably, STL) to broadcast systems (such as a timing-only eLoran or improved WWVB) to fiber-based systems.

In the final section of this chapter, we summarize four federal initiatives that are relatively low cost and that also provide some robustness for those decisionmakers still concerned.

1. Information-sharing and standards
2. Expansion of law enforcement authorities
3. Timing-only backup
4. Expansions of economically successful but geographically limited systems.

Impact of Worldview

Different audiences and decisionmakers can view the world in quite different ways. Some focus on specific threats or natural hazards while some focus on specific applications for PNT. These views reflect real, if particular, concerns and thus must be perceptible in our analysis for the analysis to be broadly understood and accepted.

For example, one oft-discussed natural hazard to GPS is an extreme space weather event, often likened to the Carrington event of 1859, which was an extremely large CME from the sun. These concerns are not limited to the United States; for example, the EU organized a meeting in late 2016 to focus on this threat (Krausmann et al., 2016). That meeting noted the potential risks to GPS and other GNSS but focused more on the need for planning to handle potentially large damage to power grids. Other reports are more dramatic, emphasizing the potential for widespread, lasting effects. For example, an article in *The Atlantic* quoted Norwegian geophysicist Pal Brekke as follows:

> With satellites and the chips inside them getting smaller as technology progresses, "one particle from the sun that penetrates a satellite can ruin things," Brekke says. "It wouldn't take that large of an event to take out all GPS." (Glass, 2016)

In contrast, an unpublished but publicly disclosed FEMA report from 2010—*Mitigation Strategies for FEMA Command, Control, and Communications During and After a Solar Superstorm*—found that, in such an event, there was a "possible" loss of enough GPS satellites to reduce the constellation below the 24 usually required, a less-dramatic and less-consequential failure (Emerson, 2017). As our earlier analysis indicates, we do not share the perception of a great risk to the GPS satellites from such an event.[1]

Obviously, there is a wide range of perceptions about the risks from such an event. Our analysis has placed disruption from a large solar storm in the context of other risks, both natural and man-made, and estimated these risks in a common economic framework that allows us to provide a cost/benefit calculation. Nonetheless, we must provide a view of our analysis that is limited to just that one hazard for decisionmakers or public audiences who are deeply concerned with its effects on GPS.

However, these decisionmakers must not neglect the other effects of an extreme space weather event. Some hazards can affect some but not all backups: for example, GNSS systems similar to GPS are very likely to be similarly affected by a solar event while most terrestrial backups would not be directly affected. Still, this hazard can yield large, non-PNT effects. In the case of a CME, the greatest effects likely would be on electrical power where the induced currents on long transmission lines can burn out expensive and difficult-to-replace transformers if the grid operators are not prepared for the effect. As the authors of the EU study argued, societal attention and investments should first resolve that problem (Krausmann et al., 2016).

The same is true of other views of the importance of the particular hazards or threats that we have addressed. For example, the overall risk estimation in a white paper from the Resilient Navigation and Timing Foundation (2016) found that "the greatest danger to the United States is from jamming." Consequently, for audiences that agree with that view, this work provides a focus on jamming—both inadvertent and deliberate—and considers the costs and benefits in that light. As our analysis indicates, we do not find the economic effects of jamming to be large enough to justify large-scale investments in alternatives, given the resiliency measures already in place in many sectors of the economy.[2]

The same is true for those concerned with a particular application or industry. For example, as noted earlier, the authors of an economic analysis of disruptions focused on the United Kingdom asserted that the greatest economic effects would be from disruptions in navigation, causing increased road congestion and delays in emergency services. In stark contrast, other commentators emphasized the criticality of timing to the financial, energy, and communications sectors of the economy (Fernholz, 2017). Although we believe our economic approach is the best way to integrate these quite different viewpoints, we also must respect the views of various audiences that might focus on a single service or, indeed, on a single use, such as communica-

---

[1] In contrast, the authors do see the potential damage to other systems, notably the power grid, as arguably large and serious.

[2] We note that this work does not apply to a commercial network of emitters, which would interfere with the GPS signals from a nearby band. The analysis in this report does not apply to such large-scale and ongoing interference.

tions or electrical power, arguing that the use is critical to all else and thus deserves special attention.

Inverse Problem: Assumptions on Threats

One way to understand the implications of our analysis is to ask, "What must one believe about specific threats or hazards to the GPS system to justify an investment to make the system more resilient and robust?" From the results in Chapter Six, we can make some clear observations.

Given the small area covered by feasible numbers of terrestrial jammers, it is implausible to expect such jamming attacks to justify any investment in costly alternative PNT systems. In part, this is driven simply by the need to provide backups everywhere to make them effective as a counter against this threat—across the nation or, in some cases, across urban areas—so the attackers have no obvious target. In contrast, more modest expenditures in such resilience-increasing elements as integrity monitoring or proliferating direction finders to responders at the local level could address incidents and help to deter them in the future, if desired. Fundamentally, however, for local attacks, a score of consequential attacks per decade would be needed to match the costs of the cheaper alternative PNT systems, and several hundred attacks would be needed to justify the expensive systems. Given the absence of such attacks to date, such investments should seem unreasonable to virtually all audiences.

Turning to nationwide events and to timing, nationwide disruptions would need to average about one day per year, primarily affecting certain sectors, such as power and communications. Such disruptions could motivate the deployment of a relatively cheap system to provide backup timing—for example, eLoran (timing only) or some enhancement of WWVB, or alternatively, White Rabbit via fiber. Because these sectors already have some in-place backups, it is difficult to distinguish White Rabbit in particular from an improved backup because it can be provided site by site, unlike a broadcast system. Of course, as noted in the threat analyses, if these disruptions are intentional, a single system dependent on a small number of terrestrial sites such as most wide area broadcast systems would offer an attractive target. Further, blowing up towers and associated equipment is much more familiar to terrorist groups than jamming an RF signal.

For nationwide events and for positioning and thus navigation as well, our analysis shows that one must believe that damages would be much greater than our estimates, counting only the damages that could be mitigated by some available system. Producing greater damages appears to require both an acceptance of the high end of the economic estimates that could practically be mitigated—perhaps a very large consumer benefit or great transportation disruptions—*and* an expectation that these significant disruptions would happen much more often, last much longer than we find credible, or both. Such beliefs are needed to justify the costs of the relevant systems to address these threats—typically such systems as MBS and Locata—because the costs of such systems are high if they need to cover many city-sized areas.[3]

In both of these cases, such high rates of nationwide disruption would be much higher than the observed rate of solar storms affecting the earth. Something else would be needed to justify a backup system. The most plausible alternative that we have found, which we discussed in Chapter Three, is a cyberattack by a sophisticated nation-state.[4]

A cyberattack on a military system (such as GPS) would likely require the resources of an advanced nation-state. To be effective, the attack would need to affect all opposing GNSS constellations available to the United States. If the attack came from Russia, we would assume that GLONASS would be shut down for the duration of the conflict, but at least Galileo would need to be disrupted as well. Because the extent and persistence of effects from cyberattacks are highly variable, such an attack could provide the number of days without PNT services that would be needed to justify a backup.

Still, there are complications. An attack on multiple GNSSs would be difficult to accomplish, even for nation-states. Additionally, if the attackers intended to harm the U.S. economy, it is hard to argue that any of the alternatives considered—commercial systems less hardened against cyberattack than GPS and the other GNSS—would not also be affected. Finally, one must assume that the two nation-states most likely to be capable of such an attack—Russia and China—would coordinate the shutting down of their national GNSS to harm the United States. Otherwise, Russia and China would need to attack each other's constellations, which would be a dangerous choice for either nation. Such coordination would be possible, although it is hard to estimate its likelihood. To find this a compelling rationale for a backup, one must believe that such an attack is possible and that it would be limited to just GNSS and so not affect a backup. We do not find this rationale compelling.

Potential Federal Actions

As our analysis makes clear, it is difficult to make a convincing argument for investing in a backup or alternative system because of the nature of these threats, independent of who is making the investment. Under most assumptions, the societal risk that could be

---

[3] As noted earlier, if the great damage is because of perceived effects at a limited number of locations—for example, disruption of the four large container ports—then the costs of mitigating the damage would also be much less, as might the appropriate federal role.

[4] Physical attacks on the GPS satellites are both difficult because of their orbital location and also because of the great redundancy of the constellation. Adding Galileo in as well would require scores of satellites to be negated before significantly reducing PNT capabilities. This makes GNSS, in general, a very unattractive target in any conflict, particularly when compared with surveillance, intelligence, or communications constellations.

mitigated does not justify the investment. Still, there are four potential federal initiatives that *do* appear to be cost-effective or close to cost-effective.

1. Information sharing and standards for hardening or quality monitoring for those systems now relying on GPS
2. Expansion of law enforcement authorities focused on GPS jamming
3. Timing-only backup through fiber/FirstNet, eLoran, or STL
4. Expansions of economically successful but geographically limited systems offering high performance (whether 5G, MBS, or other) in neglected areas. Such expansions would be a potential opportunity for a PPPs or a service-level agreement.[5]

We discuss each below.

Information Sharing

Information sharing between the federal government, which is more aware of potential disruptions in GPS, and various user communities could be important. For some users, this would primarily motivate careful system engineering, particularly among those who might be unusually sensitive to GPS disruption or spoofing. One such example concerns the PMUs used to synchronize the electrical power grid, which, some argue, could be spoofed into making the grid unstable and causing damage (Humphreys, 2012). This, of course, assumes that the PMU is not doing integrity checks on the GPS time signal. As Humphreys notes, the potential damage could be greatly reduced by hardening the systems in a variety of ways. Because the costs of hardening and the potential damages would, in this case, largely be felt by the same economic actor, information sharing should motivate appropriate responses. In cases in which some damages would be felt by others, such as the rate payers of a utility absorbing the costs, regulations requiring such hardening become appropriate.

Expansion of Law Enforcement Authorities

In a sense, a variation of hardening would allow a *societal*-level hardening to reduce the prevalence of nuisance jamming. Currently, only the FCC is allowed to levy fines against those who interfere with GPS, thus making it very unlikely that individuals using jammers for personal reasons will be caught and fined. Changing the law to allow state and local authorities to levy fines could allow for deploying such systems as "JammerCam," which is available in the United Kingdom (Chronos Technology, 2018). This appears to be a low-cost (or even profitable) alternative that could reduce such drive-by jamming wherever it becomes bothersome, although we would

---

[5] Both public-private partnerships and service-level agreements involving PNT are of particular interest to Congress, as shown in Sec. 1618 of the 2017 National Defense Authorization Act.

not expect such law enforcement to become a major activity. We also recognize that the necessary changes in laws make this a longer-term alternative, at best.

National Timing-Only Backup

The third initiative is a timing-only backup system, presumably motivated by the loss of an accurate GNSS signal during a large CME. Notably, the U.S. government has long operated radio signals for timing, such as WWV and WWVB, through NIST. This makes timing a particularly familiar federal role.

A timing backup system appears to be important for the wireless communication sector and for the electrical grid.[6] Although existing 4G cellular systems would be able to function for weeks without GPS and would continue to be available as 5G is deployed, the small 5G cells in wireless systems might not function without GPS. Timing could be important if the 5G cellular infrastructure becomes an important PNT alternative—for example, for emergency services within structures.

The degree to which the timing synchronization of the 5G architecture will rely on GPS—or, more generally, on GNSS—is a choice that has yet to be made. The network might be made to provide its own synchronization, in part to avoid the costs of a GNSS receiver for each small 5G cell (Li et al., 2017). In particular, the network's ability to internally pass timing accurately enough would also be needed to use FirstNet or any publicly supplied fiber timing signal, thus making national timing-only options redundant to the existing cellular system of clocks.

For an eLoran-type solution applied to 5G, the cost of the receivers would likely be even more of an issue than for GPS because the eLoran receivers are unlikely to be less expensive than the current GNSS chip sets. If STL were used for a timing backup, the cost of STL receivers would at best be comparable to the cost of GPS receivers, and this cost could be a barrier for 5G adoption, just as the cost is a barrier for GPS.

The electrical grid is in some sense simpler. Here, the costs of another timing signal, as for the PMUs, need to be compared with the use of existing holdover clocks for the few days per decade that they would likely be needed. Information-sharing on potential GPS disruptions with the grid operators alone could motivate internal investments in better backups.

Coverage Expansion of High-Performance Complements to GPS Through Public-Private Partnerships

There is an opportunity for a PPP to expand the scale of systems that are otherwise being deployed in limited areas. It appears likely that some system—5G, MBS, or something else—will be deployed in metropolitan areas, thus providing more precise position location of and for cell phone users.[7] The federal government could allow the

---

[6] As noted, the financial sector, which also uses timing from GPS, has high-reliability backups.

[7] The general approach and use of a PPP is provided in Appendix D.

marketplace to determine the winning solution and leverage that into solutions that are tailored to address a subset of users of national importance that might not be sufficiently profitable to attract purely market-based solutions.

Initial deployments of such new systems are naturally targeted to the areas where they offer the highest economic payoff. These areas tend to be higher-income areas, for example, because this is where the ads sent to cell phone users are more valued. If the U.S. government considers entering a PPP specifically to provide PNT, it would have to carefully evaluate the best way to proceed, given rapid changes in the commercial arena and other programs that would change the context of any further intervention, such as FirstNet.

Implications of the FirstNet PPP for PNT

FirstNet was established by Congress in 2012 to provide a nationwide, high-speed broadband network for the public safety community, including law enforcement, fire, and emergency medical personnel. It received $6.5 billion in initial funding through an FCC spectrum auction and 20 MHz of federally owned spectrum to host the network. The legislation requires the network to be self-financing and allows FirstNet to collect user or subscription fees from public safety entities and from "a public-private arrangement to construct, manage, and operate the nationwide public safety broadband network" in exchange for access to network capacity on a secondary basis for non-public-safety services.[8]

Although a PPP is not specifically mandated in the legislation, FirstNet's governing board determined that it would be the best approach to finance the network. First-Net issued a request for proposals (RFP) in January 2016 and awarded a 25-year contract to AT&T in March 2017. Under that contract, AT&T will invest $40 billion to build, operate, deploy, and maintain the network. In exchange, AT&T will be able to use the spectrum for commercial purposes when it is not being used by first responders and will be able to charge user fees to first responder agencies (these fees must be competitive with other cell phone service providers, however, because agencies are not mandated to use the network).

FirstNet had a staff of 30–35 contracting, legal, and technical subject-matter experts to develop the RFP and run the acquisition process. It also hired another 30–40 people who were responsible for conducting outreach with the states and first responder agencies because the legislation required each state to decide whether to "opt in" to the nationwide network or build its own state network, based on AT&T's proposed network coverage in each state. This suggests that a staff of less than but on the order of 100 people would be needed for a new, complex PPP for a backup PNT system.

---

[8]   Public Law 112-96 (2012). This section is also based on FirstNet: First Responder Network Authority (undated) and discussions with FirstNet personnel.

One of the requirements of the PPP contract is that AT&T provide "z-axis" coverage in major metropolitan areas with dense vertical structures, which would help determine the vertical location of a first responder in a multistory building. The geographic extent of this requirement is not yet well defined but would most likely apply to the 20–100 metropolitan areas with the highest-density populations. One possible technology is the NextNav MBS, but it will probably be another 12–18 months before AT&T makes a firm decision on a z-axis solution. The solution would then have to go through the procurement and deployment processes before being fielded.

Because the coverage area is not yet settled, it is unclear how much of the United States would be covered. To give some context for estimating this, the curve in Figure 7.1 shows an upper bound in the economic activity that might be covered, assuming that all the MSAs below a certain size have this deployment. The MSAs are ordered by size, from the largest (New York/Newark/Jersey City) down to the smallest (Sebring, Florida). Presumably, this ordering also, very roughly, tracks the priority for deploying the FirstNet z-axis capability, as larger MSAs are more likely to have high-rise buildings that drive this need. Although not all of an MSA would have high-rise buildings requiring z-axis coverage, the existing MBS deployment in the San Francisco Bay area, a reasonable case involving about 100 transmitters, covers not just the built-up areas but also the entire East Bay (Oakland, Berkeley, etc.) and south to San Jose within the rimming mountains, thus capturing much of two MSAs. This also makes sense for deployment because AT&T presumably would seek to sell this service more broadly.

Figure 7.1
Cumulative Gross Domestic Product in the Largest Metropolitan Statistical Areas

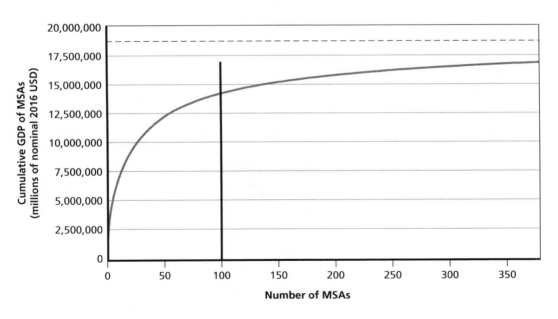

In Figure 7.1, the dashed line represents the total U.S. GDP in that year. The vertical line, at 100 MSAs, shows the economic value contributed by the 100 largest MSAs. The comparison with the dashed line shows the dominance of urban areas in the U.S. economy.

The key point illustrated in Figure 7.1 is that a FirstNet deployment of something like NextNav, backing up GPS for essentially all of the economic benefits from PNT within its coverage area, would be likely to capture a great deal of all the benefits available to any backup simply because so much economic activity is focused in urban areas. This is because this deployment can *independently* provide positioning with comparable accuracy to GPS and good timing. No other alternative would be needed in these areas. A similar implication would almost certainly be true even if AT&T were to choose a different z-axis solution.

Of course, areas outside these urban areas would not gain these benefits from this FirstNet z-axis solution, so such sectors as agriculture and mining would still suffer whatever losses might be incurred in a GPS disruption. But, as we have seen, no systems other than GNSS-based ones cover large areas with the accuracy required. For small areas such as mines or ports, local deployments are possible but seem likely to be decided on and funded on a case-by-case basis.

This FirstNet deployment also offers an opportunity for the federal government. Once a decision is made on a z-axis solution, DHS could potentially leverage FirstNet's contract with AT&T to expand the system to still more areas as a PNT backup. The contract is structured such that additional task orders and funding could be added by other federal agencies. This approach could avoid the need for new congressional action and reduce the size of the staff needed to implement a new PPP, although sufficient staff would be needed to discuss options with the various sectors using PNT and to justify where expanded backup services would be warranted.

However, we do not believe that any single system, including expanding whatever z-axis solution is chosen, will be sufficient for all users. For example, users requiring higher-positioning accuracy will still need to procure a system providing such accuracy, such as Locata. Fortunately, market solutions will handle many of the challenges identified in this report. Public provisioning is not necessary for many economically significant users with concentrated interests and the ability to pay, such as banks or telecommunications firms. The PPP based on FirstNet seems most likely to be useful in covering some of the leftover gaps after the urban centers are covered.

Final Observations

Our analysis shows that large expenditures for deploying a backup or complementary PNT system are difficult to justify in an economic sense. One must believe in very large damages, frequent disruptions, or widespread effects, for which we found no evidence. PPPs or service-level agreements do not affect this calculation as those arrangements would merely change the source of the funding.

The federal government may still choose to support some deployment for noneconomic reasons. In those cases, we have outlined four options that would be relatively low cost and would also offer at least modest benefits.

Nationwide Framework for Valuing the Change in Productivity Enabled by GPS

In the analysis supporting the nationwide portion of Chapter Four, the valuation of changes in GPS availability is based on standard economic theory. Many applications of GPS enable a sector of the economy to produce a given level of output at lower cost; that is, they lead to an increase in productivity. In this section, we discuss how to value such an increase in productivity as well as some potential pitfalls in doing so.

We measure the value of a productivity increase by the resulting increase in consumer plus producer surplus. Figure A.1 illustrates the concepts of consumer and producer surplus in a supply and demand diagram.

The diagram shows that equilibrium price (P^*) and quantity (Q^*) are determined where supply equals demand. The demand curve shows the willingness to pay of consumers for any unit of output, and the difference between willingness to pay and price is a measure of the welfare consumers derive from participating in the market. The difference is called "consumer surplus," and it is measured by the area ($A\alpha P^*$). The supply curve shows the marginal cost of any unit of output, and the difference between price

Figure A.1
Illustration of Consumer and Producer Surplus

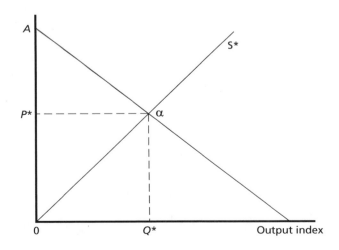

and marginal cost is the benefit that producers derive from participation in the market. The difference is called "producer surplus," and it is measured by the area ($P^*\alpha0$). The sum of consumer and producer surplus (area [$A\alpha0$]) is then the economic welfare that society derives from this market.

Figure A.2 shows how the effect of a technological improvement on economic welfare is calculated in this framework.

Technological progress in any sector is represented by an outward shift in the supply curve, illustrated here by a shift from supply curve $0S^*$ to curve $0S'$. The new equilibrium price (P') and quantity (Q') are determined where the new supply curve crosses the demand curve. Consumer surplus increases from area ($A\alpha P^*$) to area ($A\beta P'$), and producer surplus (which may increase or decrease) changes from area ($P^*\alpha0$) to area ($P'\beta0$). The sum of consumer and producer surplus—that is, the economic welfare that society derives from this market—increases from area ($A\alpha0$) to area ($A\beta0$).

We characterized the impact of technological change in Figure 5.2 as "an outward shift in the supply curve" and illustrated it with this specific case as a clockwise rotation. We now discuss more formally how any specific kind of technical change would be represented by a shift in the supply curve. There is no obvious way to represent the "level of technology" in the supply curve; it is simply a functional relation between P and Q. We will abstract in this discussion to linear forms of both the supply and demand curves. Since we are analyzing the impact of *changes* in P and Q from initial equilibrium levels P^* and Q^*, we are only concerned with the behavior of the supply and demand curves in the vicinity of the initial equilibrium. Given this, the

Figure A.2
The Effect of a Technological Improvement on
Economic Welfare

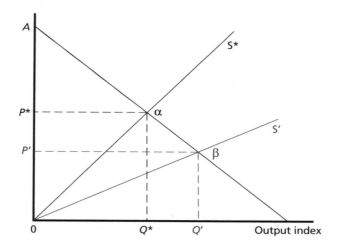

linear functional forms we are using can be interpreted as linear approximations[1] of the actual supply and demand curves around the initial equilibrium point (P^*, Q^*). We will define the linear demand curve as

$$P = A + BQ,$$

($B < 0$), and the linear supply curve as

$$P = g + hQ,$$

($h < 0$). Equilibrium P and Q are thus

$$Q = \frac{(A - g)}{h - B}$$

$$P = \frac{(hA - bG)}{h - B}.$$

With this functional form of the supply curve, we can now more explicitly characterize the impact of technological change. In economic theory, the supply curve is derived from an underlying *production function*, in which the level of output depends on both the level of inputs (such as labor, capital services, and intermediate goods) and the level of technology. We can generally write this as

$$Q = f(I, \theta)$$

Here I is an index of intermediate inputs and θ is an index of technology. The level of output that will be produced at any price (i.e., the supply function) is at the point on the production function at which profits are maximized. This is at the point at which the marginal product of each input (i.e., the derivative of the production function with respect to that input) times the price of output equals the unit cost of the input. As represented in the production function shown above, we will abstract to one composite input index, defined so that its price is unity. Then the profit maximization problem is

maximize $PQ - I$

subject to $Q = f(I, \theta)$.

And the supply curve is determined by the point at which

$$P \frac{dQ}{dI} = 1$$

---

[1] More formally, these are first-order Taylor series approximations around the equilibrium points (P^*, Q^*). B and h are first derivatives calculated from the elasticity values.

or

$$P\frac{dI}{dQ}.$$

A linear supply curve is derived in this way from a quadratic production function of this form

$$I = gQ + \frac{h}{2}Q^2 + \xi$$

since

$$P = \frac{dI}{dQ} = g + hQ.$$

Alternately, we can interpret the quadratic production function as a quadratic approximation to the actual production function since it leads to a linear approximation of the supply curve.[2]

We now can represent technical change in terms of how the production function changes, consistent with economic theory. We will characterize technical change in this way (which is, as we will see, consistent with how it is characterized in the literature on the impacts of GPS on economic welfare): We define productivity in terms of Q/I, that is, output per unit input. We define a z percent increase in technology as an increase in Q/I by the factor $\gamma = 1 + (z/100)$; in the case of a decline in technology, z would be negative. For example, we define a 10-percent increase in technology as an increase of Q/I by the factor 1.1. The precise definition is that a z percent increase in technology means that any output level, say Q_0, which was produced by input I_0 before the technical change, can be produced after it by input I_1, where I_1 is determined by

$$\frac{Q_0}{I_1} = \gamma \frac{Q_0}{I_0}$$

or

$$I_1 = (I_0 / \gamma).$$

Now we represent the production function before the technical change by the equation

$$I_1 = g_0 Q + \frac{h_0}{2}Q^2 + \xi_0.$$

---

2   Again, these are Taylor series approximations.

We define a z-percent improvement in technology as a shift in the production function to

$$I = g_1 Q + \frac{h_1}{2} Q^2 + \xi_1,$$

where

$$g_1 = \frac{g_0}{\gamma}$$

$$h_1 = \frac{h_0}{\gamma}$$

$$\xi_1 = \frac{\xi_0}{\gamma}$$

$$\gamma = \left(1 + \frac{z}{100}\right).$$

This then implies that the supply curve shifts from

$$P = g_0 + h_0 Q$$

to

$$P = g_1 + h_1 Q,$$

where

$$g_1 = \frac{g_0}{\gamma}$$

and

$$h_1 = \frac{h_0}{\gamma}.$$

We now have an exact way to represent the impact of a z-percent improvement in technology in our linear supply and demand framework. We note that other ways of characterizing a z-percent change in technology could have been defined and that non-linear functional forms could have been used. We judge that our approach is appropriate for this application.[3]

---

[3]   Additional sensitivity analysis on these assumptions would be a valuable extension of this analysis.

An Aside on Deriving Supply and Demand Curves from Supply and Demand Elasticities

The responsiveness of supply and demand to price is generally characterized by the *elasticity of supply* and the *elasticity of demand* rather than by the slopes of the curves. The demand elasticity is the percentage change in the quantity demanded when price increases 1 percent, defined as

$$\eta = \frac{dQ}{dP}\frac{P}{Q} = B\frac{P}{Q}$$

(note $\eta < 0$). Similarly, the supply elasticity is the percentage change in the quantity supplied when price increases 1 percent, defined as

$$\varepsilon = \frac{dQ}{dP}\frac{P}{Q} = h\frac{P}{Q}$$

(note $\varepsilon < 0$).

In this work, we will generate our linear supply and demand curves by specifying (1) that they result in a base quantity and price for the sector, designated as Q^* and P^*, and (2) that they have a specified elasticity at that point. The base quantity and price are generally the current market values, and from there we can estimate the economic cost of losses of GPS services. Given the elasticities, we can derive the parameters of the demand and supply curves as

$$B = \eta\frac{Q^*}{P^*}$$
$$A = P^* - BQ^*$$
$$h = \varepsilon\frac{Q^*}{P^*}$$
$$g = P^* = hQ^*.$$

The Impact of Supply and Demand Elasticities on the Economic Welfare Effects of Technological Change

We are concerned in this report with the economic impact of a loss of GPS services, which means losing the productivity gains associated with using those services. Therefore, we will continue the discussion in terms of a productivity decrease. The economic impact of such a decrease in technology depends on the supply and demand elasticities prevailing in the market affected. We will illustrate this in this section. Throughout, we will define (P^*, Q^*) as the price and output prevailing in the market with availability of GPS, and (P', Q') as the price and output prevailing in the market after loss of availability of GPS.

One way to assess the economic impact of such a decrease in technology due to loss of GPS services is the following. First assess the deterioration in production technology that results from losing GPS services (say this is y percent; $y < 100$). Then assess that in the absence of GPS services the price in the market would rise by the factor $\varphi = [1 - (1 - (y/100))]$; that is, $P' = \varphi (P^*)$. Then, assess that output in the market will not change when GPS services are lost; that is, that $Q' = Q^*$. Finally, calculate the welfare loss as $(P' - P^*) Q^*$.

We now show that this is equivalent to doing the consumer and producer surplus calculation with (1) $z = (-y)$, (2) zero elasticity of demand (i.e., the demand curve is $Q = Q^*$) and (3) infinite elasticity of supply (i.e., $h^* = h' = 0$, so price is independent of output). The impact on economic welfare of a change in technology under these assumptions is illustrated in Figure A.3. For simplicity, we have used $P^* = Q^* = 1$ in this example.

Figure A.3 shows that the result of the technical decline is indeed for the price to increase by 25 percent, which is $1/\gamma$, and the change in welfare is the area with width one and height 0.25, as in the word example above.

A zero demand elasticity and an infinite-supply elasticity certainly could obtain; that is, there is nothing in economic theory that makes them inadmissible. However,

Figure A.3
Impact of a 20-Percent Technical Decrease in a Market with Demand Elasticity Zero and Supply Elasticity Infinite

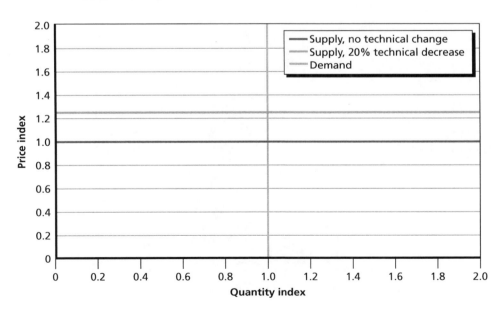

we judge that there are good reasons to believe that they are not representative of what the effect of loss of GPS for some period of time would be. Purchasers of output can be expected to have some response to a higher price, so a nonzero demand elasticity is reasonable. Suppliers can be expected to have constraints on how quickly they can change production levels while infinite-supply elasticity implies that any level of output demanded will be supplied at the fixed price of g (the supply curve is $P = g$ at infinite elasticity). Demand elasticity can range from zero to minus infinity, and supply elasticity from zero to plus infinity. Unity is an attractive "representative" value for demand elasticity since the weighted average of *all* demand elasticities, weighted by their budget share, has to be unity given a budget constraint on buyers. There is no such natural supply elasticity, but infinity is not plausible in a relatively brief period for the reasons given above, and zero is not plausible for any activity whose scale can change as economic incentives change. We will use a value of unity as a representative intermediate level in our next example.

Figure A.4 shows the effect on welfare of a 20-percent technical decline if demand elasticity is minus one and supply elasticity is one.

Figure A.4 illustrates that in this case market price rises to 1.11, and market quantity falls to 0.89. The new level of economic welfare is the area of the triangle with base 2 and height 0.89 (the new market quantity), or 0.89, so the welfare loss is 0.11. Thus, with these elasticities, the economic welfare loss associated with the technical decrease is about half that of assuming zero demand elasticity and infinite-supply elasticity.

Figure A.4
Impact of a 20-Percent Technical Decrease in a Market with Demand Elasticity Minus One and Supply Elasticity One

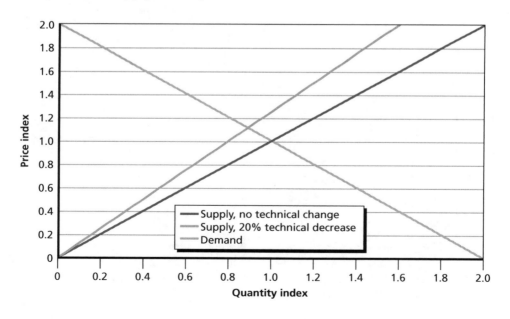

Analysis of many elasticity cases shows that welfare loss goes down as supply elasticities fall (i.e., supply becomes less elastic), and that it also goes down as demand elasticities increase in absolute value (i.e., demand becomes more elastic). The common sense of this is that (1) elastic demand implies that users of the output can find good substitutes when prices rise, and (2) inelastic supply implies (through the production function) that the relatively fixed level of production is not responsive to input level and thus not to technical changes that affect how much input is needed to produce any given level of output.

Empirical Estimates of the Cost to Society of a Loss in GPS Services

The total costs to U.S. society of a complete or partial loss of GPS services can be conceptually divided into two categories.

The first category is **economic losses.** These are the losses in economic welfare, measured by GDP, that result from loss of GPS services. There are two existing estimates of these losses to the U.S. economy.[4] The first estimate of economic losses is described in two reports: Leveson (2015a) and Leveson (2015b). The second estimate of economic losses is presented in Pham (2011). There is also one estimate of losses to the U.K. economy, presented in Sadlier et al. (2017). These three studies are broadly consistent with each other, finding that loss of GPS services would cost the economy about one half of 1 percent of its GDP.[5] In the two estimates for the U.S. economy—Leveson and Pham—the bulk of the losses are in grain production, construction-related activity (including earth moving and surveying), road freight transportation, and consumer use of GPS-enabled services, such as personal navigation. We will discuss these economic losses below.

The second category is **public safety losses.** These are losses that occur if emergency services that rely on GPS are slowed in their arrival at the site of the emergency, thus resulting in additional loss of life or property. Such emergency services include ambulance, police, and firefighting services, whose life- and property-saving effectiveness depends on prompt response; such prompt response is facilitated by GPS and would be slowed in its absence. To our knowledge there is no previous estimate of this public safety cost for the United States. We suggested a range of possible estimates in Chapter Four of this report.

---

[4]   These two previously published studies have now been superseded in many, but not all, sectors, by the generally more detailed study by O'Connor et al. (2019).

[5]   When we inflate Pham's figures to 2018 dollars as we do Leveson's, his total savings from GPS in those sectors is $51 billion compared with Leveson's $46 billion. We inflated Sadlier et al.'s results to 2018 equivalent dollars by multiplying by the ratio of U.S. to U.K. GDP at market exchange rates. Their estimate after that adjustment, net of the emergency services sector, is $49 billion.

First, we describe the work of Leveson, presented in his reports of June 2015 and August 2015, although we will modify his estimates in part; we will discuss those modifications in this section.

The second column of Table A.1 shows Leveson's estimates of the impact of GPS on economic welfare, by sector, in 2013, which is $55.7 billion—0.33 percent of the 2013 GDP of $16.7 trillion. As the table shows, more than 99 percent of the economic impact is in five sectors: grain production, earth moving, surveying, commercial road transport, and consumer location-based-services. We have projected these to 2018 values (the third column), inflating them by a factor of 1.22, which is the ratio of esti-mated 2018 GDP to 2013 GDP.[6] Doing so leads to a 2018 economic benefit of $67.7 billion. Inflating by nominal GDP growth is equivalent to assuming that the economic benefits of GPS services grow proportionately with both real output and the price level.

We make five modifications to Leveson's estimates, which we judge result in an estimate more appropriate for our purposes—namely, assessing the value of investment in GPS-backup or PNT resilience programs.

First, Leveson's estimates of benefits in the first four sectors of Table A.1 (grain, earth moving, surveying, and commercial road transport) follow the methodology that we described above: if the productivity gain from GPS is estimated to be y percent, Leveson measures economic benefit as $(P' - P^*) Q^*$, where Q^* is the current level of

Table A.1
Leveson's Estimated Annual Economic Benefit of GPS Services in 2013 and 2018 (billions of then-year dollars)

Sector	2013 Economic Benefit	2018 Economic Benefit
Grain production	13.7	16.7
Earthmoving in construction	5.0	6.1
Surveying	11.6	14.1
Commercial road transport	11.9	14.5
Consumer location-based services	13.1	15.9
Other[a]	0.4	0.5
TOTAL	55.7	67.7

SOURCE: Second column from Leveson (2015a, 2015b); third column calculated by us as described in text.

[a] "Other" includes timing services, and air, rail and maritime transportation. As discussed elsewhere, Leveson's analysis of these other sectors is incomplete and we do not use those numbers as a basis for analysis.

---

[6]   Actual GDP through 2017 is from BEA (2019). Projected 2017–2018 GDP growth is from Chairman of the Council of Economic Advisers (2018), Table 8-1.

output, P^* is currently prevailing price, $P' = \varphi\,(P^*)$, and $\varphi = [1/(1-(y/100))]$. This approach is analytically correct, given the implicitly assumed demand and supply elasticities of zero and infinity, respectively, and those elasticities certainly could obtain; that is, there is nothing in economic theory that makes them inadmissible.

We judge that there are good reasons to believe that the elasticities assumed by Leveson are not representative of what the effect of loss of GPS for some period of time would be. Purchasers of output can be expected to have some response to a higher price, so a nonzero demand elasticity is reasonable. Suppliers can be expected to have constraints on how quickly they can change production levels while infinite-supply elasticity implies that any level of output demanded will be supplied at the fixed price of g (the supply curve is $P = g$ at infinite elasticity). Demand elasticity can range from zero to minus infinity, and supply elasticity from zero to plus infinity. Unity is an attractive "representative" value for demand elasticity because the weighted average of *all* demand elasticities, weighted by their budget share, has to be unity given a budget constraint on buyers. There is no such natural supply elasticity, but infinity is not plausible in a relatively brief period for the reasons given above, and zero is not plausible for any activity whose scale can change as economic incentives change. We judge that a demand elasticity of −1 and a supply elasticity of 1 are more appropriate values for calculating economic benefits. Making this change in the analysis reduces the economic benefits of these four sectors (grain, earthmoving, surveying, and commercial road transport) by about half; the precise level depends on the size of y. Leveson's calculations of the benefits for consumer location-based services is not based on a productivity increase in production, as the first four sectors of Table A.1 are; it is based on a "willingness-to-pay" estimate. No elasticity-based adjustment is appropriate for that.

Second, Leveson's estimate of benefits in other sectors is explicitly characterized by him as incomplete; he said, "estimates were not included for some sectors because of insufficient data" (Leveson, 2015a, p. 8). Further, to the extent that Leveson did estimate benefits in the timing-related sectors of finance, telecommunications, and the electrical grid, they were based on the avoided cost of building a different nationwide timing system—this is a different kind of counterfactual calculation that does not capture the GNSS benefit that would be lost during a GNSS outage (since the different nationwide timing system is not actually in place).

We therefore supplement Leveson's estimates based on the work of Pham (2011). Pham too estimated benefits in the agriculture, surveying, and road transport sectors, with results broadly consistent with Leveson's. He then estimated benefits in other commercial sectors as 71 percent of the benefits in agriculture, surveying, and road transport based on the relative share of commercial purchases of GPS equipment in different sectors. We show Pham's estimates in Table A.2. Further, while Pham folded the "other" commercial sectors together and assumed the benefit of GPS in those sectors was proportional to GPS spending on equipment in the 2010 to 2015 period in the aggregate, we also show the results if one does not combine the sectors and instead

Table A.2
Estimates of Annual Economic Benefit of GPS Services (billions of then-year dollars)

Sector	Pham (2011 $)		Leveson (2013 $)	Extrapolated Range of Estimates (2018 $)		
				Low	Medium	High
Consumer location-based services	Not estimated		13.1	15.9	46.7	92.9
Commercial road transport	10.3	10.3	11.9	7.2	22.2	44.9
Agriculture	19.9	19.9	13.7	7.9	12.4	17.9
Earthmoving	9.2[a]	3.6	5.0	3.0	8.8	15.6
Surveying		5.6	11.6	5.2	5.6	6.1
Aviation	28.3[a,b]	2.4[b]	0.14	1.4[b]	2.3[b]	3.4[b]
Maritime		22.2[b]	0.19	13.1[b]	21.1[b]	31.1[b]
Railway		0.05[b]	0.06	0.03[b]	0.05[b]	0.07[b]
Timing		3.42[b]	0.05[c]	2.0[b]	3.3[b]	4.8[b]
People tracking		0.2[b]	Not estimated	0.1[b]	0.2[b]	0.3[b]
TOTAL	**67.7**		**55.7**	**55.9**	**122.6**	**217.0**

[a] Pham (2011) combined these sectors when estimating benefits.

[b] This estimate rests on an assumption that GPS benefits are roughly proportional to GPS spending across sectors.

[c] This estimate is based on the cost of building an alternative timing system that has not actually been built.

make the same assumption that share of GPS benefit is proportional to share of GPS equipment spending sector-by-sector.

We call this the "Low" estimate in Table A.2 because we make two other adjustments to the estimates, both of which increase them. First, Leveson provided an estimate of GPS penetration in each of the markets, and he noted that it has grown over time. We adjusted for assumed continued penetration growth by increasing the penetration level from P, Leveson's level, to $P + (0.5*(100 − P))$; i.e., so that the increase in penetration closes half the gap between Leveson's estimated level and full penetration.[7] Second, Leveson also estimates the share of precision location or timing benefits that are provided by GPS, as opposed to alternate or ancillary systems. In the case of a sudden loss of GPS services, many of the benefits from the ancillary systems may also be lost; that is, they may require GPS availability to themselves function properly.

---

[7]   This is admittedly simply an assumption about how penetration may have increased; more definitive research on this, beyond the time and resource constraints of this study, would improve the estimate.

Given the wide array of alternate and ancillary systems in the several sectors, we make this overall assumption on the consequences of GPS loss: let g be Leveson's estimate of the share of precision location or timing benefits that are provided by GPS. We increase this to $g + (0.5 (100 - g)/2)$; that is, so that the increase in the GPS-provided share closes half the gap between Leveson's estimated level and a share of unity.[8] The estimate of benefits adjusted this way is shown as our "Middle" estimate in Table A.2. For the "Other" sector, we assume that the penetration and GPS-provided shares are a weighted average of those shares in the grain, earth moving, surveying, and commercial road transport sectors. The shares are based on Leveson's estimates of the economic benefits of GPS to these sectors as shown in the last column of Table A.1.

The "High" estimate in Table A.2 shows a case in which GPS penetration increases to 100 percent and in which the GPS share is unity—that is, in which all precision location and timing benefits are lost in the absence of GPS. These estimates, Low to High, bracket the range of effects of other elements of the PNT ecosystem to cushion the effects of GPS loss or disruption.

---

[8]   This too is simply an assumption about the consequences of GPS loss; more definitive research on this, beyond the time and resource constraints of this study, would improve the estimate.

Supplemental Information on the Regionally Based Microeconomic and Geographic Analysis

This appendix will speak to some of the threats and vulnerabilities of the sectors discussed in the main body of the report and provide more detail on their composition and the way in which we thought about their potential exposure to the loss of GPS. It is intended as a supplement to Chapter Four and should be viewed as rounding out the analyses contained therein.

Construction

Construction has the NAICS code 23. This sector encompasses the following NAICS subcodes (see Table B.1):

The most significant portion of this sector from a GPS jamming perspective is site preparation as it involves earth-moving equipment that can make extensive use of automated machine guidance applications. GPS can also be used in site preparation planning and the compliance assessment of those plans. There are other applications of GPS in tracking and site management, but earth-moving applications appear to bring the most significant productivity gains and cost reductions. To a lesser extent, large civil engineering projects and line installation (both power and water/sewer) use GPS, but they also routinely require confirmation from other sources.

GPS is not the sole source of positioning guidance on sites; GPS augmentation systems, micro-IMUs, GIS-based systems, building-information modeling (BIM),[1] cellular and wireless communications like Bluetooth, image recognition, robotic total stations, and vastly improved computers and control systems all play a part. These systems sit on top of a hodgepodge of construction techniques from multiple eras. At any given time, the construction industry has techniques and processes from ancient to cutting edge running side by side. Crews can adapt to changes through the application

---

[1] BIM was designed to enhance planning and allow engineers and architects direct insights in better control of planned and current construction projects.

Table B.1
North American Industry Classification System Subcodes

Code	Title
237115	New Single-Family Housing Construction (except For-Sale Builders)
236116	New Multifamily Housing Construction (except For-Sale Builders)
236117	New Housing For-Sale Builders
236118	Residential Remodelers
236210	Industrial Building Construction
236220	Commercial and Institutional Building Construction
237110	**Water and Sewer Line and Related Structures Construction**
237120	**Oil and Gas Pipeline and Related Structures Construction**
237130	**Power and Communication Line and Related Structures Construction**
237210	**Land Subdivision**
237310	**Highway, Street, and Bridge Construction**
237990	**Other Heavy and Civil Engineering Construction**
238110	**Poured Concrete Foundation and Structure Contractors**
238120	Structural Steel and Precast Concrete Contractors
238130	Framing Contractors
238140	Masonry Contractors
238150	Glass and Glazing Contractors
238160	Roofing Contractors
238170	Siding Contractors
238190	Other Foundation, Structure and Building Exterior Contractors
238210	Electrical Contractors and Other Wiring Installation Contractors
238220	Plumbing, Heating and Air-Conditioning Contractors
238290	Other Building Equipment Contractors
238310	Drywall and Insulation Contractors
238320	Painting and Wall Covering Contractors
238330	Flooring Contractors
238340	Tile and Terrazzo Contractors
238350	Finish Carpentry Contractors
238390	Other Building Finishing Contractors
238910	**Site Preparation Contractors**
238990	All Other Specialty Trade Contractors

NOTE: Bold type indicates a field that uses GPS.

of hybrid techniques and field-expedient measures, permitting construction activity to continue even in the face of unexpected hiccups.

The actual process of construction as we view it consists of all the activities listed above. The job site consists of a set of processes that can be thought of processes with differing degrees of independence from other part of the process. Some cannot start until other processes have completed while other processes run in parallel. The larger the site, the more elements the project manager must sequence. In large projects, managers can stagger operations to allow for slack in the event of a delay in one section of the project without an automatic critical path failure.

However, even though GPS is jammed at a construction site, one cannot assume that the site has shut down. Because only a small number of activities on a construction site actually require GPS to function in any significant manner, construction may be less exposed to GPS outages and interference than might be assumed. The impacts of GPS outages vary based on the construction project and the activities going on at a given time. While GPS and automated machine guidance is finding its way into many activities, at a given site different aspects of the construction project have significantly varying levels of exposure. Here we can highlight the differences by discussing the trades where GPS can help in terms of increasing productivity.

Any reader familiar with any construction project, from a simple home renovation to the largest construction projects in the world, would instantly recognize that not every portion nor even an overwhelming majority of construction tasks utilize or require GPS to function. Even a temporary work stoppage is not necessarily a loss of much money as weather stoppages happen on construction sites as well. Employers may send employees home early, ask them to return later to work longer hours, or shift activities. Apart from permitting, equipment rental, or other fees, there is minimal long-term impact on their income.

Surveying

One of the areas which has benefited the most from the widespread proliferation of GPS has been the surveying industry. This sector encompasses the NAICS codes 541360, or Geophysical Surveying and Mapping Services, as well as 541370, Surveying and Mapping (except Geophysical) Services. The 2012 Economic Census reported annual receipts of $3.14 billion ($3.47 billion in 2018 dollars) in geophysical surveying and $5.59 billion ($6.18 billion in 2018 dollars) for other kinds of surveying and mapping.

The Census Bureau estimated that the receipts reported by these two areas was roughly $8.75 billion nominal 2012 dollars. It was deemed unlikely that GPS was either uniformly used or uniformly vulnerable across these two sectors. GPS used in Geophysical Surveying and Mapping tends to be used in pit mining. Mines are typically located far apart from one another and in mountainous areas. This leads to the use of alternatives to GPS at the mines themselves and would necessitate a jammer placed at each individual mine to have an appreciable impact. Surveying and Mapping (excluding Geophysical) teams are more commonly observed in day-to-day life and more visible and thus more likely to be targeted.

Surveying (of the kind considered in this report) is a combination of office and fieldwork. In the case of boundary surveying, or the survey of an area that was previously examined or which contains dig hazards, the job begins with a records search. This may result in rough boundary lines or hazard maps and may also provide survey monuments-markings that serve to indicate particular locations or coordinates. Once this has been collected, a surveyor (or team) will go to the area. A small, regular section can be examined relatively quickly. A large, irregular, or never-before-surveyed area will take longer. Once on site, the survey team will determine reference points based on the records. These may come from GPS coordinates or preexisting monuments. Once these reference points have been determined, the rest of the plot can be surveyed using prisms and photogrammetry or GPS units, usually in the course of a day. More importantly, while GPS coordinates may be used for an initial check of the location of buried lines, nonetheless, a trained surveyor must go out to the site with rods and other tools to locate the "as built" rather than "as planned" locations of pipes. The GPS units are faster when there is a clear line of sight to the satellite and the reference point. However, in areas that lack that clear line of sight, photogrammetry is still the preferred method. While a GPS unit has the added benefit of a reduction in team size, some total robotic survey stations can also be operated solo, meaning that efficiency is not the sole purview of GPS. Once the area has been fully measured and marked, the survey team must draft documents for the client. This is a much longer process lasting about a week. Once it is complete, it must be reconciled with the previous records. Where there are discrepancies, the survey team must determine the cause.

The majority of surveying is not fieldwork, and the majority of fieldwork is not GPS dependent. Surveying involves using GPS or cueing off of "monuments" with known locations. Monument-cued surveying may take slightly longer than GPS-based surveying, but it is a substantial requirement for licensing new surveyors and tends to be an important element of ensuring property boundaries are correctly mapped. So long as surveyors are trained to examine indoor or underground locations, the knowledge of non-GPS surveying will continue.

Perhaps the most visible fieldwork of a surveyor is the work performed on construction sites, often before other work has begun. Unlike construction work, survey fieldwork has a very short period of vulnerable (if any) to GPS jamming. The surveyor need only find a monument or receive one definite GPS coordinate to be able to determine everything else on site. While the non-GPS surveying may be slightly slower than pure GPS work, it is nonetheless a fairly rapid process on site, with most of the work stacked on the front and back ends with desk work.

Agriculture

One of the industries to benefit most from the widespread proliferation of GPS is the agricultural sector and specifically, wheat farming. This sector encompasses the NAICS codes 1111, or Oilseed and Grain Farming, which can include soybeans, corn, wheat. The agricultural sector has benefited from the integration of GPS receivers into many of their activities, including farm planning, and tractor guidance, among other uses (National Coordination Office for Space-Based Positioning, Navigation, and Timing, 2018). In many ways, the agricultural sector represents an ideal environment for GPS jammers, as farm land is often flat, and wide open, which allows for maximal reach. Unfortunately for the would-be attacker, many farms are involved in activities on a daily basis that are not particularly sensitive to access to GPS coordinates. Field laborers harvesting crops like grapes and strawberries by hand are not using GPS to do so, while crops that can be harvested by machine enjoy benefits from doing so. On the other hand, the number of days per year which represent a critical agricultural period which could be targeted by adversaries is relatively small. In order to estimate the potential economic loss through the denial of GPS to agriculture, the focus must shift from metropolitan areas toward more rural counties.

Farmland in America is not distributed in an even fashion geographically or economically, and GPS usage in farming is similarly uneven in distribution of value. GPS driven planters allow farmers to avoid over-planting fields to save seed and fuel (Staff, 2010). Estimated savings in 2009 put a GPS unit as providing a 2–7 percent cost savings per acre. Given GPS unit prices at the time, depending on the actual cost savings, the crops involved and several other factors, farmers could expect to break even somewhere between 200 and 2,100 acres of crop planting (Groover and Grisso, 2009). As of 2012, 55 percent of farmland was held by only 4 percent of farmers, while nearly 70 percent of farmers operate farms of 179 acres or fewer (USDA, 2014). The break-even cost does not justify the use of GPS units for many farmers.

However, this does not account for costs and savings associated with targeted spraying of pesticides and herbicides. Per another study, the break-even rate for herbicide and pesticide spraying depends heavily on the number of different kinds of pests and weeds, and how much of the field must be sprayed. If an infestation is caught early enough, targeted spraying works, and can save money and labor. However, if the infestation has or might have spread, more of the field must be sprayed in order to prevent additional outbreaks. Spraying any more than 75 percent of a field means that targeted spraying is pointless from a cost-savings perspective (Adamchuk et al., 2008). In addition to the spraying itself, the field must be surveyed, and weeds and pests must be identified. This hinges on low-cost field surveying, which is not available in all locations, and/or machine recognition of targets.

GPS signals only matter during a handful of days—tilling and planting, the targeted application of sprays, and the harvest. These events do not fall on the same day across the country; corn harvests, for example, typically occur in a cascading fashion starting in July and ending in December. In the case of tilling and planting, there is some leeway to plant later or to use non-GPS-enabled planters. It may be less efficient, but the job will still get done. A schedule slip is not usually a huge problem on the front end. In the case of the application of herbicides and pesticides, GPS failure means less efficiency and higher costs as well as the externalities associated with increased chemical usage. However, spray applicators can still function, and their jobs will still get done. Harvesting is the real choke point. Once a crop is ready, it should be harvested as quickly as possible. Farmers can use harvesters without GPS to continue the harvest, albeit with slightly less efficiency. The harvest days will be longer than intended, but the work can still get done. Even there, however, there is some leeway. Silage crops can stay in the field a bit longer. Grains that will need to dry in a grain elevator can stay in the field for a day or two than originally intended, without complete failure. If rains come, then the ears of wheat will lose their robustness and kernels will swell and split. However, if the weather is reasonably dry, the farmer can continue to harvest at a slightly slower pace. The window of vulnerability to GPS jamming is limited on farms, and even then, most work can continue, albeit at a slower pace.

Finance

One of the areas which has taken advantage of GPS for timing purposes is the finance sector. This sector encompasses the NAICS codes 521 (Monetary Authorities-Central Bank), 522 (Credited Intermediation and Related Activities), and 523 (Securities, Commodity Contracts, and Other Financial Investments).

Many financial transactions are reliant on PTP or NTP. Those reliant on PTP often use GPS as a form of confirmation of timing. Rapid trades depend on knowing who held which stocks at any given time. When trades occur in the course of milliseconds, this knowledge depends heavily on precision timing. GPS currently serves as a simple way to get that precision timing.

Per a white paper on timing in financial trading, GPS-based timing is "subject to disturbances." Financial firms should account for such disturbances and can generally survive such outages if they make the appropriate investments in backup systems and networks. This could mean using an atomic clock or local oscillators or having multiple points of access to GPS satellites should allow a given firm to weather unexpected outages. Some white papers suggest that an atomic clock can provide precision timing to the necessary level for weeks or months, more than enough time to outlast a jammer. These backup features tend to be expensive, but given the expected losses associated with the temporary loss of PTP when there is no backup, they are cost-effective.

In estimating the impact of GPS jammers on the finance industry we focused on the sector in a single metropolitan statistical area with large amounts of financial services. Los Angeles has a large financial sector. We took the sector GDP of finance located solely in Los Angeles and then looked at a sample of major financial institutions in the region. The majority are geographically clustered in the downtown area. Moreover, most of the money is concentrated in a very small number of firms. The remainder is scattered in smaller firms around the area. Major firms, by virtue of the amount of money they handle, invest heavily in backup systems. While an atomic clock is expensive for most individual users, a financial firm working with billions of dollars might see it as a reasonable investment. As of 2017, the New York Stock Exchange was already working with the National Physics Laboratory to use their atomic clock for precise timing. Smaller organizations are less able to invest in expensive backups. The impact of GPS jamming is thus understood to be the impact of jamming checked against the probability of backups failing, by the amount of money processed by any given firm at any given day, averaged out across the year. Jamming when all backups function is meaningless. The largest firms handle the most money on a daily basis while having a lower probability of a backup failing. Smaller firms handle less money on a daily basis but have a higher probability of a backup failing, so the overall losses might, in fact, be higher in this group. The overall losses are determined by adding up the expected value of all of these losses.

Technical Overview of GPS Enhancements and Alternatives for Degraded GPS Environments

Overview of Enhancements and Alternatives

In this appendix, we provide a description and technical assessment of GPS enhancements and alternatives that are potential candidates for use in one or more critical infrastructure sectors. We use a common outline for each option discussed in the following subsections. Each subsection provides a description of the technology and the quality of PNT that it can achieve, and discusses implications for new infrastructure and/or new user equipment. In considering performance, we do not require all alternatives to perform as well as GPS since some critical infrastructure sectors do not require accuracy as good as GPS.

Wireless PNT Signals

Multi-GNSS Chip
Description and Performance
Even in benign environments, it is advantageous for receivers to exploit signals from multiple GNSS constellations. By receiving a larger number of signals from which receivers can compute pseudorange, users can (1) attain somewhat higher PNT accuracy, (2) assure PNT in unfavorable environments, such as urban canyons, and (3) improve signal monitoring by checking for inconsistent values of pseudorange.[1] If all four major constellations could be processed on one chip, then well over 100 signals would be available globally. Table C.1 summarizes the available number of signals in each of the bands used by GNSS constellations.

Because of these advantages, many GPS receivers exploit two or more GNSS signals already. A 2016 survey of GPS receiver manufacturers found that nearly 65 percent of their designs exploit at least two constellations, with GPS and GLONASS being the most common among the various possible combinations. Most smartphones

---

[1]   Such signal monitoring is called Receiver Autonomous Integrity Monitoring.

Table C.1
Approximate Planned Number of Signals in Each GNSS Band

System	L1/L1OC/E1/B1	L5/L5OC/E5a/B2a	L2/L2C/L2OC	E6/LEX
GPS	30	30	30	Not used
GLONASS	24	24	24	Not used
Galileo	30	30	Not used	30
BeiDou	35	35	Not used	35
QZSS	3	3	3	3
IRNSS	Not used	7	Not used	Not used

SOURCE: European Global Navigation Satellite Systems Agency, *GNSS User Technology Report*, Luxembourg: Publications Office of the European Union, No. 1, 2016.

built within the last several years already support this capability. Figure C.1 shows the results of this survey. As Galileo and BeiDou become operational, it is very likely that the number of receivers exploiting those signals also will increase.

The majority of nonmilitary, consumer-oriented receivers today use only the L1/E1/B1 frequency band. However, Broadcom has released a mass-market GNSS chip, the BCM477755 chip, that simultaneously receives GPS L1 and L5, Galileo E1 and E5a, QZSS L1 and L5, GLONASS L1, and BeiDou B1 and achieves 30-centimeter accuracy (Moore, 2017; Broadcom, undated). The first smartphone with this chip, the Xiaomi Mi 8, became commercially available in May 2018.

Figure C.1
Percentage of Receivers Supporting Multiple GNSS Constellations

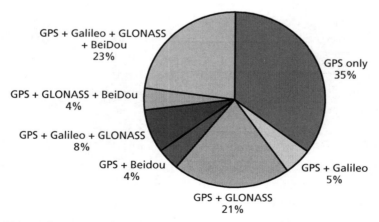

SOURCE: Data from European Global Navigation Satellite Systems Agency, *GNSS User Technology Report*, Luxembourg: Publications Office of the European Union, No. 1, 2016.

If GPS signals should be degraded due to some event specific to the GPS con-stellation or due to spoofing that mimics the unique structure of the GPS signal, then multi-GNSS receivers may maintain good PNT, with only a slight degradation of per-formance. However, in the case of deliberate noise jamming, an attacker could likely jam all GNSS signals due to their shared frequency band and weak signal levels.

Implications for Government and User Investments

While using multi-GNSS chips to improve PNT performance in the event of degraded GPS signals requires new user equipment, the implied requirement for investment is mitigated by the fact that chip manufactures are investing in this technology and con-sumers are buying it already. Users do not need to be persuaded of the risks to GPS to invest in this technology; they already are doing this for the advantages of multi-GNSS chips in everyday situations where GPS is operating as expected.

Enhanced Long-Range Navigation
Description and Performance

Enhanced Long-Range Navigation (eLoran) was designed to be a modernized and improved version of the now-defunct Loran-C system. Loran-C was a timing and radionavigation service intended to provide positioning accurate to within about 0.25 nautical miles, or 460 m (Narins, 2014). That level of accuracy was useful to mariners, although it was insufficient for harbor navigation. The error in repeatability of the position calculation and relative location with respect to nearby users could be several times better and is useful for the relative navigation and safety of ships (USCG, 1992). Loran-C was used in a variety of applications; however, the market for receivers dimin-ished over time as Loran-C was supplanted by GPS and other GNSS. The Loran-C network was retired in the United States in 2010. eLoran was to make use of the same transmitter sites and much of the existing Loran-C infrastructure. As of 2016, the only eLoran broadcast in North America was the 360 kW signal from the former USCG Loran Support Unit transmitting site in Wildwood, New Jersey.

eLoran, like its Loran-C predecessor, broadcasts in the 90–110 kHz band, which enables its signals to propagate without suffering from the line-of-sight limitations that are typical of most other signals designed for PNT. At this low frequency, eLoran sig-nals can penetrate into indoor environments. Signals from eLoran transmission sites are synchronized and coded so that user receivers can compute the time difference of arrival for any pair of signals. Synchronization is maintained by three atomic clocks at every transmitting station.[2] Any pair of transmitters is sufficient to localize the posi-tion of the receiver on a hyperbola. For this reason, Loran is sometimes called a hyper-bolic navigation system. Three transmitters are sufficient to uniquely allow the receiver to find its 2D (horizontal) position and time (much as four GPS satellites are necessary

---

[2] The redundancy in atomic clocks improves overall timing accuracy and improves holdover duration in the event of a GPS outage.

for users to find their 3D position and time).[3] However, a single transmitter is sufficient to provide time transfer to a receiver if the receiver's location is already known.

The ground-wave radio signals suffer propagation delays due to the environment, which in turn introduce time and position errors if the delays are not perfectly known. The propagation delays are divided into the Primary Factor (PF) due to atmosphere, the Secondary Factor (SF) due to radio propagation over the ocean surface, and the Additional Secondary Factor (ASF) due to the electrical properties of terrain other than seawater. Of these, the ASF is the most difficult to model (Blazyk et al., 2008). In particular, changes in conductivity due to soil moisture are a major source of uncertainty in the ground-wave propagation speed. (In addition, propagation by skywave paths that enable receivers to see signals that arrive by multiple reflections off the ionosphere can cause ambiguity in the time delay measurement. This latter source of uncertainty, however, is substantially mitigated through signal encoding so that receivers can select the first arrival of a transmitted signal and not a skywave "echo.")

Although the uncertainties arising from ASFs are not predictable, the errors can be measured at reference stations to allow for better PNT performance if those corrections are communicated to users. The exploitation of such corrections is called differential Loran, which is well understood, but was not widely implemented within the United States (Carroll et al., 2004). Due to the spatial correlations of the ASFs and their slow variations over time, one reference station can support precision eLoran within a distance of about 35 miles. Multiple reference stations could support a model of ASFs that could support users over a larger area.

eLoran is meant to have significantly better performance than Loran-C (Helwig et al., 2011). Two improvements are related to timing accuracy. First, eLoran specifies tighter tolerances on waveforms. Second, it requires that each transmitter be more tightly synchronized to UTC. Another important improvement is the addition of a data channel, sometimes called the Loran Data Channel, which can carry useful information such as differential eLoran corrections, differential GPS corrections, eLoran or GPS integrity data, date and time, emergency messages, or other information useful to users. eLoran without differential correction was expected to provide time transfer with better than one μs accuracy and with differential correction to better than 100 ns. This quality of performance has now been experimentally demonstrated (Offermans, Bartlett, and Schue, 2017; UrsaNav Inc. and Harris Corporation, 2017). Differential corrections should allow positioning accuracy of about 20 meters, and one model suggests that accuracy better than 10 meters may be possible (Safar, Vejrazka, and Williams, 2011).

---

[3] In fact, Loran transmitters are configured in groups called "chains," with one transmitter in each chain being designated as the "master" and others as "secondary" transmitters. Groups typically have three to six transmitters. More than three are useful to avoid circumstances were one pair provides unfavorable geometry for the hyperbolic solution. Some transmitters were configured to support more than one group.

The GLA of the United Kingdom and Ireland pursued an implementation of an eLoran network with the hoped-for cooperation of other European nations. The GLA produced a map of ASF in U.K. coastal waters, which however did not capture temporal variation in ASF. Using that map, GLA found that with maritime navigation by eLoran, within 30–50 meter accuracy was possible (Ward et al., 2015). However, the governments of France, Norway, Germany, and Denmark decided that the benefits of the eLoran network did not justify the cost and shut down their transmitters, ultimately leading the GLA to abandon the project also (Saul, 2016). The eLoran tower in Anthorn, Cumbria, is broadcasting until the contract runs to term, but there are no known current users of the towers (Proctor interview, 2018).

Saudi Arabia has upgraded its five Loran stations to be eLoran-capable. India is also reportedly interested in upgrading its six Loran stations and perhaps expanding the network (Narins, 2014). South Korea is pursuing an eLoran system for use in port entry, motivated by concerns over North Korean GPS jamming. South Korea is currently testing the capabilities and has not reached initial operational capability but aims for an FOC in 2019 (Seo interview, 2018).

A current eLoran receiver board is 60×30×8 mm (Reelektronika, 2017), not including the antenna. Past Loran receivers have typically been larger and used a whip antenna at least a few feet long. However, antennas as small as two square inches may be sufficient for eLoran (Gallagher, 2017). Further miniaturization may be possible. Thus, while it remains unclear whether eLoran antennas could be made small enough for integration into smartphones, integrating eLoran receivers into systems used to time-stamp ATM, credit card, and high-speed market transactions applications in which GPS is now the main source of timing definitely would be feasible (Glass, 2016). In 2016 UrsaNav demonstrated the application of eLoran for providing precision timing to the New York Stock Exchange. Exterior antennas were not needed due to eLoran's low-frequency signals being able to penetrate into buildings (Schue, 2016).

In the absence of GPS, the synchronization of eLoran transmitters could degrade, although holdover capability is potentially very good. Any impact from absence of GPS would not result in severe underperformance for a considerable time. For example, a well-designed ensemble of atomic clocks could provide holdover with accuracy remaining better than 50 ns for up to 70 days (Dahlen, 2008). eLoran transmitters might also get time directly from UTC standards at NIST or the USNO by space-based or terrestrial links to avoid any direct dependence on GPS.

Implications for Government and User Investments

The adoption of eLoran as a national alternative to GPS would require acquisition of both infrastructure and user equipment. Infrastructure can leverage legacy Loran-C transmitter sites and control stations if they are still available. Unfortunately, the dismantling that occurred in 2014–2015 will increase the cost of reuse of legacy infrastructure. Infrastructure investment might also include the establishment of reference sites to support precision applications and to provide PNT performance that would be somewhat inferior to existing GPS services. Additionally, eLoran sites, if they are the primary backup to GPS, would require fairly stringent physical and cybersecurity investments because of their exposure on the ground and concentration at a few sites.

Users will require new receiver equipment. Although eLoran is compatible with legacy Loran-C receivers, users of legacy receivers will not be able to access the key benefit of eLoran—namely, the information carried on the data channel to allow for precision applications. For eLoran to serve as a functional backup to GPS with a performance that is only somewhat inferior, most users would require new equipment. eLoran receivers could potentially be integrated into GPS equipment for fixed and larger mobile users as was done with at least some legacy Loran-C user equipment.

One possible use of eLoran would be to provide time transfer nationally. Offermans, Bartlett, and Schue (2017) calculated that only four transmitting sites would be needed to support all of CONUS with 1 μs accuracy. A notional laydown of 71 reference sites to support modeling of ASFs could support differential eLoran and precision timing to fixed users with 100 ns accuracy and would cover nearly all major cities, airports, and ports within CONUS. Twenty or more transmitters would be needed to provide position and navigation services for all of CONUS and to provide accurate timing to mobile users.

Pseudolites and Beacons

In the following several subsections, we discuss various pseudolite and related technologies. In this discussion, we use the terms pseudolite and beacon to mean any terrestrial transmitter that broadcasts a signal designed for user positioning and navigation, which is typically received by user equipment that computes the pseudorange to transmitters whose locations are known.[4] The term "pseudorange" refers specifically to an estimated range obtained by multiplying the speed of light by the time the coded signal takes to travel from transmitter to receiver (Rao and Falco, 2012). Pseudorange differs from true range because the propagation time, measured at the receiver, contains contributions from multiple error sources including receiver clock errors, transmitter clock biases, variations in the speed of the carrier wave due to passage through media of varying densities (e.g., the troposphere and ionosphere), the effects of multipath, and errors in the known position of the transmitter or satellite (Van Sickle, 2018).

---

[4]   We note that Raquet (2013) reserves the term "pseudolite" for signals that are similar to GNSS and uses the term "beacon" for signals that are substantively different. While the underlying principle is essentially the same in both cases, we also differentiate systems to avoid confusion.

Modified GPS Signals

Description and performance

It is possible to design pseudolites that are entirely compatible with existing GNSS user equipment, broadcasting on the same carrier frequency using a signal structure that is the same as GPS or any other GNSS. A key motivation for such systems is for indoor PNT where space-based signals are not available. Several architectures have been proposed. In one case, transmitters surrounding a building repeat and amplify GPS signals that are acquired by an outdoor antenna. User receivers will compute pseudoranges that are incorrect due to the extra path length from transmitter to user, but if the user knows the location of the transmitters, he can then postprocess the data to find his correct location (Xu et al., 2015). Another architecture uses transmitters that broadcast GPS signals and ephemeris data corresponding to simulated satellite orbits. Users again will compute pseudoranges that are wrong, but postprocessing based on knowledge of the transmitter locations results in an accurate position (Kim et al., 2014). At least some pseudolite concepts might potentially be implemented at relatively low cost using SDRs and other commercially available equipment.

While these pseudolite concepts are attractive because they achieve GNSS-quality PNT (Kim et al., 2014) using existing user equipment, there are some significant concerns. Most significantly, they could interfere with reception of actual GNSS signals. By design, pseudolite signals will be much stronger than actual GNSS signals since they are intended to be received where authentic GNSS signals do not penetrate. Even if pseudolites were operated only as a backup when GPS was unavailable, pseudolites still suffer from the so-called "near-far" problem, in which some signals can be much stronger than others due to terrestrial geometry, again making detection and processing of all pseudolite signals potentially difficult. An advantage of pseudolite signal strength is the substantial jamming resistance that results.

The above difficulties may not be insurmountable, and it seems possible that architectures that allow both GPS and pseudolite signals to operate simultaneously may be possible. However, there is no such architecture that is generally accepted and no pseudolite architecture based on the GPS/GNSS carrier frequency that has been commercialized. Indeed, it would be illegal to broadcast signals that interfere with GPS in the United States. Any pseudolite architecture would have to prove it is compatible with GPS before it could be entertained as a viable backup option, which would be a very difficult threshold to attain.

Implications for government and user investments

Pseudolite architectures based on GNSS signals would require infrastructure and modest modifications to user equipment. Generally, four or more transmitters would be needed for any service area.

User equipment will require, in addition to the existing GNSS capability, communications and postprocessing to correct for the locations of the pseudolite transmitters. This could be provided easily by any smartphone with a cellular connection, although this would introduce a secondary dependence on cellular communications.[5]

Locata

Description and performance

The Locata Corporation, a private Australian company with a U.S. subsidiary, has developed a system of terrestrial beacons to provide precise PNT signals to specially equipped receivers in a localized area. Able to provide centimeter-level positioning precision and under one ns timing synchronization between transmitters without the use of atomic clocks, this system can serve as an augmentation to GPS or as a complete alternative in areas where the beacons have been deployed. Locata was originally developed in 1997 as an alternate local GPS signal, using the same L1 frequency as GPS itself. In 2005 the system was evolved to use a proprietary signal structure carried on the license-free Industry Scientific and Medical (ISM) band, popularly known as the Wi-Fi band (2.4–2.4835 GHz) (Rizos, Gambale, and Lilly, 2013). The system consists of transmitting beacons (called "LocataLites") which transmit on two ISM frequencies, modulated with pseudorandom noise number codes. The original intent was to develop a system that provides centimeter-level position accuracy in locations where GPS signals could be obstructed or degraded, such as deep valleys, open-cut mines, forested areas, and urban and indoor environments (Rizos, Gambale, and Lilly, 2013).

One of Locata's key innovations is its patented "TimeLoc" technology, a method by which it can establish and maintain synchronization between LocataLites to within one ns of one another without the need for atomic clocks in each transceiver. In contrast, the GPS system utilizes atomic clocks on each of its satellites, updated regularly via ground stations from the USNO master clock, to maintain synchronization. Locata's TimeLoc procedure allows each LocataLite to repeatedly adjust its own transmissions to match the timing of signals received from other LocataLites, the only requirement being that each beacon is able to receive signals from at least one other (Powers and Colina, 2015b). This enables the beacons to "cascade" the synchronization across all the beacons in the "LocataNet." The differences in reception time due to geometric position of the LocataLites are taken into account during this process, since the precise locations of each LocataLite are determined (via geodetic surveying and/or GPS methods) when they are originally installed, and those precise LocataLite positions are broadcast in each LocataNet signal transmission. The TimeLoc procedure is described in U.S. Patent 7,616,682 (Small and Locata Corp., 2009). Figure C.2 shows the core components of every Locata implementation, a LocataLite transceiver, and a "Rover" receiver.

While the TimeLoc procedure enables highly precise synchronization among transmitters, the LocataNet time can drift relative to UTC unless a UTC source or

---

[5]   Our assumption is that transmitter locations cannot be transmitted in the GPS message. This is simply due to the current GPS message structure, not due to any physics-based limitation.

Figure C.2
Core Components of a Locata Implementation

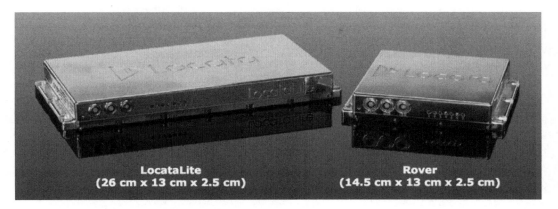

LocataLite
(26 cm x 13 cm x 2.5 cm)

Rover
(14.5 cm x 13 cm x 2.5 cm)

SOURCE: Product image courtesy of Locata.
NOTE: A Generation 4 LocataLite transceiver (left), and a RV8 "Rover" Locata receiver (right).
Additional antennas and power components are required for an operational LocataNet.

GPS time is fed into the network as a master clock. Thus, any applications that require accurate UTC would require such a master clock input to the LocataNet. Without such an input, the LocataNet still provides the same centimeter-level quality of precise positioning (Powers and Colina, 2015b).

In 2013, the University of New South Wales independently demonstrated Locata time transfer within about five ns over distances up to 45.4 miles (Gauthier et al., 2013). In 2015, the USNO conducted independent testing of LocataLites in the District of Columbia urban environment, with transceivers separated by as much as 1.8 miles, and demonstrated their ability to maintain synchronization within 0.2 ns (Hsu, 2015).[6] Also, while there is no theoretical limit to the number of LocataLites that could be synchronized in a LocataNet using TimeLoc, the USNO concluded from their experiments that, in practice, each additional intermediate LocataLite introduced approximately 25 picoseconds of "jitter" into the synchronization (Powers and Colina, 2015a).

There are several examples of existing Locata deployments in various facilities around the world. One is the Locata implementation at an open pit gold mine in Australia (Rizos, Gambale, and Lilly, 2013). In 2012, Leica Geosystems deployed a 12-LocataLite network at the Newmont Boddington Gold Mine in Western Australia to augment the existing GPS-dependent guidance systems used for vehicle fleet management inside the open-cut mining pit, which at the time was 1 km long × 600 meters wide × 275 meters deep. The GPS-only systems had been experiencing reduced accuracy near the pit walls as the pit grew deeper. Leica designed receivers

---

[6] For comparison, GPS is required to maintain synchronization to at ≤20 ns 68 percent of the time (DoD, 2008).

called the Jigsaw Positioning System (JPS) that combined inputs from GPS satellites and Locata ISM transmissions and installed them on drilling rigs for position and heading purposes near the pit walls. The system reportedly resulted in an improvement of drilling machine availability time between 6.5 percent and 23.4 percent, depending on the location of the machines in the pit, which translated into cost savings of approximately $112,700 over two months for two drills since drill downtime costs the mine approximately $1,000 per hour per drill. The accumulations of cost savings over the years have reportedly been even more significant as all 11 drills have had new receivers installed.

Another example of a Locata deployment is at the White Sands Missile Test Range in New Mexico (Craig et al., 2012). The 746th Test Squadron (746 TS) at Holloman Air Force Base is chartered to provide test and evaluation of guidance and navigation warfare for the U.S. Department of Defense. In 2010, the 746 TS began developing the Ultra High Accuracy Reference System (UHARS) to provide improved position and velocity accuracies "up to seven times better" than its previous truth system, the Central Inertial and GPS Test Facility Reference System. One component of the UHARS is the non-GPS-based positioning system (NGBPS), needed to provide submeter position accuracy for airborne and land-based test vehicles in a GPS-denied jamming environment. The 746 TS selected Locata to provide this capability, funding a contract in 2010 to first enable Locata to upgrade its systems' capabilities to meet the requirements for the NGBPS, which were more advanced than what Locata had demonstrated in its previous commercial deployments up to that point. In particular, Locata needed to enhance its capabilities for longer range acquisition and tracking (30 miles), higher-power transmission, and different antenna designs as well as the ability to tolerate dynamic aircraft conditions for its receivers while still providing better than 18 centimeters position accuracy (accounting for position dilution of precision) for aircraft traveling at approximately 550 mph. Locata demonstrated these capabilities in October 2011 at the NGBPS Tech Demo at White Sands Missile Range (WSMR) with a network of ten LocataLites. Aircraft with Locata receivers acquired the signal at an average range of 48.8 miles and with tracking starting at about 38 miles. RMS carrier solutions of 15 cm in height and six cm horizontally and a 3D RMS of 17.4 cm were demonstrated, thus meeting the 18-cm 3D RMS requirement. The U.S. Air Force declared initial operating capability of the UHARS in August 2016 with a LocataNet of 16 LocataLites attached to transportable, solar powered trailers distributed over a 20×20 square mile area on the WSMR North range (Kawecki et al., 2016).

The differences in Locata deployment in the two examples discussed here illustrate that LocataLite configurations are designed to suit specific user requirements. LocataNets also have been installed in warehouses, airports, and seaports to back up GPS and/or improve available precision, especially for logistics applications.

Because Locata is a completely independent PNT system from GPS, it represents a localized complete PNT backup with good performance in GPS degraded environment and the potential to provide indoor signal availability as well as improved positioning accuracy and timing synchronization. The system would not be affected by GPS jammers since it utilizes a different portion of the spectrum, though in principle Wi-Fi jammers could be targeted against Locata receivers. If a critical infrastructure application required accurate UTC, then Locata would have to be augmented with an input from one of the official UTC master clocks or GPS. This is the only potential dependence of Locata on GPS availability.

Implications for government and user investments

Use of Locata requires deployment of a LocataNet comprised of LocataLite transceivers and antennas as well as specialized Locata user receiver equipment. While companies like Leica (mentioned above) have created integrated JPS user devices that act as receivers for both Locata and GPS, there is no standardized Locata microprocessor that has been integrated with consumer-level user devices (as is happening with multi-GNSS chips and the NextNav MBS system discussed below). Such development may be possible, but at present, deployment of a LocataNet generally requires users to purchase all parts of the system. To date, these systems have been installed at local industrial or government facilities for PNT backup of GPS or for users requiring centimeter-level precision.

Metropolitan Beacon System

Description and performance

The NextNav LLC has developed a system of terrestrial beacons to provide precise PNT signals to mobile device users. Branded as MBS, it can serve as a complementary PNT source where GPS signals are unreliable (e.g., urban canyons) or provide primary PNT capability where GPS signals are too weak to penetrate (e.g., indoors). The MBS beacons use a five-foot vertical omnidirectional antenna to radiate up to 30 W of transmitted power in the 920–928 MHz band. MBS is FCC certified with a part 90 license transmitter that allows transmission at up to 30 W. At this power level, NextNav transmitters can be separated by up to 10 km in rural areas and about 0.5 km in dense urban areas (Johnson, 2016). NextNav claims that this allows them to use fewer beacons than some competing technologies.

MBS has been installed in several metropolitan areas for demonstration. It creates a signal that is much more powerful than GPS satellite signals at ground level (which is typically approximately 30 dB below the noise floor). Therefore, the significant processing power and time for acquisition by the user equipment needed for GPS (receiver equipment running high current for between 30 seconds and 12 minutes depending on availability of assistance information from cell towers) are not needed for MBS utilization due to the higher power of the MBS signal in the deployed region. This allows an acquisition time of one to six seconds instead, with lower power draw on the user device (Gates and Pattabiraman, 2016). Also, unlike cellular positioning using LTE cell towers, the MBS system does not require usage of the LTE wireless spectrum to operate, since the beacons are independent of LTE cell towers (GPS World Staff, 2016).

NextNav claims the system can provide under 10-ns synchronization using a single beacon as long as the user is stationary at a known location. The system currently uses GPS as a source for UTC. Each MBS beacon has its own atomic clock, and common-view GPS time transfer is used to synchronize between beacons within 2.5 ns (Johnson, 2016; van Graas and Meiyappan, 2014; NIST, 2016). If GPS is unavailable, the rubidium atomic clocks in each beacon provide holdover to maintain synchronization between beacons, though the system will eventually drift from UTC over time. NextNav has claimed to be exploring development of two-way time transfer between beacons as well as synchronization with USNO as a direct source for UTC, which would eliminate the dependence of MBS on GPS in the future (van Graas and Meiyappan, 2014).

Users must receive signals from at least three transmitters to determine a 2D horizontal position via standard trilateration techniques, which can be combined with a differential pressure measurement to provide vertical position as well (Johnson, 2016). In principle, MBS can use multilateration with four or more beacons to provide true 3D positioning, but vertical accuracy based on ranging signals is typically poor for terrestrial systems due to the limited vertical separation of the beacons. NextNav's use of differential pressure measurements between the beacons and user devices allows users to determine altitude with a demonstrated median of one to three meters (or "floor level") vertical accuracy for indoor positioning (NextNav, 2013). They claim five-to-ten-meter horizontal accuracy throughout the metropolitan regions where the system has been installed, which includes a "commercial-grade" deployment in the San Francisco Bay area (over 900 square miles between Marin County and San Jose) and "initial builds" in 39 other metropolitan regions as of October 2013 (GPS World Staff, 2013).

While the initial deployment of the beacon system itself progresses, the user equipment required to use MBS is being adopted into telecommunications standards for mobile devices which would mean there is a large installed base of user equipment where the MBS is deployed. In October 2013, Broadcom acquired commercial licensing to NextNav's MBS technology, enabling the company to integrate MBS support into mass-market connectivity and mobility platforms used in cell phones and tablets (GPS World Staff, 2013). The 3rd Generation Partnership Project (3GPP) is a group of seven telecom standard development organizations that make decisions about what telecom protocols are standardized into upcoming generations of mobile processors. The 3GPP Release 13, released in 2016, included general messaging support for Terrestrial Beacon System location technologies, including MBS (GPS World Staff, 2016). This means that support for MBS will become available for any Release-13 (or later) compliant LTE network-capable user device globally, similar to the previously standardized GNSS systems, allowing for a transparent user experience. Theoretically, the mobile phone user should be able to transfer seamlessly from GPS support to MBS support, for example, without needing to take any action.

The ability of MBS to provide positioning in urban environments and indoors, with floor-level accuracy, is also a significant benefit for the E911 system, which automatically provides the location of callers to 911 dispatchers. Indoor location provision from devices dependent only on GPS is typically coarse by comparison (Next-Nav, 2015). The MBS system was evaluated for E911 location accuracy by a working group of the FCC's Communications Safety, Reliability and Interoperability Council (CSRIC) in 2013. In tests in San Francisco, the council measured MBS indoor average vertical errors of between 0.6 and 3.6 meters (depending on local building density, ranging from "rural" to "dense urban") and average horizontal errors between 27 and 70 meters. Note that CSRIC describes an E911 standard of horizontal accuracy within 50 meters as providing a meaningful indoor location. NextNav's errors were the lowest of the three technologies tested in this indoor accuracy evaluation (CSIRC and Working Group 3, 2013; Cameron, 2013).

Because MBS is a PNT system completely independent from GPS, it represents increased PNT reliability for mobile devices with standard compatible processors when in an instrument-deployed region. The system would not be affected by GPS jammers since it utilizes a different portion of the spectrum, and the more powerful signal at ground level would make it more difficult to jam than GPS signals.

Implications for government and user investments

Adopting NextNav's MBS as a backup for GPS has implications for acquisition for both infrastructure and users. Infrastructure costs will depend on the scale of system deployment. Currently, MBS has only been deployed at a commercial scale in the San Francisco Bay area, though there are initial deployments in at least 39 other metropolitan regions of the United States. Considerable expansion would be required for Next-Nav to provide a nationwide PNT alternative.

User equipment that could ultimately prove to have minimal cost to most consumers is being developed. Much like standardized adoption of support for multi-GNSS chips, mobile processors capable of supporting the MBS are also being standardized into mobile devices today. NextNav appears to be on a path that could enable widespread commercial adoption. Any additional cost would be the minimal differential of using GNSS/MBS chipsets, which could be produced in vast quantities, instead of GNSS chipsets. It seems possible that more expensive receivers could be developed for some users with high-precision requirements, much as there are now such GNSS receivers, but we have not seen any proposals for such user equipment.

Signals of Opportunity

SoOPs are defined as RF signals that are not intended for navigation (Raquet, 2013). These can include such things as AM/FM radio, digital TV, Wi-Fi (IEEE 802.11),

Bluetooth, radio-frequency identification (RFID), and cellular telephony. Advantages of SoOPs are that (1) they are ubiquitous; (2) many are high power compared with GPS, whose signal is typically below system noise levels and therefore would be difficult to jam; (3) many can penetrate buildings or are available in indoor environments; and (4) common user devices like smartphones already receive many of these sorts of signals. Disadvantages of using such signals for PNT are that (1) since they were not designed for PNT, resulting PNT accuracy is likely to be worse than GPS; (2) positioning is possible only if transmitter locations are known; and (3) such signals will not be available everywhere.

One can imagine a future situation in which personal devices like smartphones and tablets share data about their RF environment with a "navigation cloud." The navigation cloud would continually update and improve its knowledge about transmitter locations, which could be provided back to users as an augmentation or alternative to GPS. This might present user privacy concerns, but is technically plausible.

SDRs could be key to such a concept since they could potentially make the fullest use of all such signals. An SDR that can exploit RF signals from 50 MHz to 2.2 GHz is already available for Android phones at a cost of only about $10 (SDR Touch, undated). This device must be connected to an antenna, which will determine which frequencies can be exploited. The external antenna might also be a limiting factor for some users. The point is that the technology is not likely to be expensive.

Below we consider three specific SoOPs that might be most useful as alternatives to GPS in the near term: Wi-Fi, LTE (4G) cellular service, and 5G cellular service. We address these due to their widespread availability and potential performance. In reality, user equipment that exploits SoOPs will probably integrate multiple signals, including GNSS, to achieve the best PNT solution in any environment. One ambitious program to develop such a capability is Aerospace Corporation's Sextant program to develop an open-source model that uses multiple SoOP signals (e.g., Wi-Fi, cellular, television, and radio) and PNT signals (e.g., GPS, eLoran, and augmentation systems), interprets them into a common data format, and then produces the best possible PNT solution (Masunaga, 2018). Efforts like this could improve PNT solutions, to include the case when GPS is unavailable, across many users and environments.

Wi-Fi/WLAN
Description and performance
Wi-Fi signals are based on the IEEE 802.11 standard. WLANs are commonly based on Wi-Fi signals. Such WLANs are ubiquitous; many homes and commercial establishments have them. Our smartphones detect them and their identifying names, even when the networks are secure and cannot otherwise be accessed. Detection ranges of such signals vary from a few tens of meters for simple home installations to perhaps many times more for more powerful and extended networks as one might find at a hotel.

Ranging that is accurate to three to four meters is possible.[7] This quality of geolocation is possible if the phone can access a database of historical Wi-Fi detections. Phones augment Wi-Fi detections with signal strength measurements to achieve the stated accuracy. In fact, smartphones routinely employ such algorithms, especially to enable faster acquisition of GPS, which otherwise could take many minutes. Such capability was first implemented in the Apple iPhone in 2008, exploiting a Wi-Fi database developed by a company called Skyhook Wireless, but later transitioned to a database maintained by crowdsourced data pulled from user phones (Fleishman, 2011). Android phones have a similar capability and exploit a database based on data collected by Google's Street View trucks. More recently, Apple purchased a company called WiFiSLAM, which claims the ability to provide location[8] to 2.5 meters (Empson, 2013).

While Wi-Fi appears to offer great potential as an alternative to GPS for positioning and navigation where such signals are common, it will not provide time transfer.

Implications for government and user investments

Since a vast number of user devices can already see Wi-Fi signals, many users might not require more than a new "app" on a smartphone. The main concern for infrastructure development, say for government users, is to ensure that the data within the "navigation cloud" is available. One issue is ensuring the availability of a connection by cellular or wired link to the "navigation cloud." A related issue is that at present, these databases appear to be proprietary to Apple and Google. There might be a role for the government to develop and maintain a navigation cloud as a public service, much as it now supports GPS as a free public service.[9]

LTE (4G) Cellular
Description and performance

LTE is a specific implementation of 4G cellular telephony that is in common use in U.S. markets. According to International Telecommunications Union (ITU) standards, LTE base stations should remain synchronized with UTC within ±1.5 µs (Goode, 2018), which is easily maintained when base stations have access to GPS signals. This degree of synchronization allows users to infer location data from LTE signals.

---

[7]   Reported in Koivisto et al. (2017). It is unclear whether such accuracy is based on measurements of signal strength, some exploitation of the signal structure, or both.

[8]   Location accuracy depends on speed of motion and on the availability of ancillary sensors, such as the compass and accelerometer in the phone. The 2.5-meter value appears to be what can be achieved at slow or nonmoving speeds.

[9]   Later we discuss the potential for a navigation cloud to also contain other sorts of data that could support SLAM methods. The case for a publicly maintained navigation cloud could be compelling since it could be an enabler of multiple PNT alternatives and be a valuable asset even when GPS is available. For example, the database conceivably could include data on GPS signal integrity, differential GPS and differential eLoran corrections and other location-specific information. It also could include data on other RF signals of opportunity, such as Apple's iBeacon which uses a Bluetooth signal (Betters, 2013).

The method employed by user equipment, like cell phones, to infer location from LTE is based on the time difference of arrival (TDOA) of each pair of signals.[10] Any pair of base stations leads to a potential location solution along a hyperbola, with the two base stations as the foci. Two or more pairs allows a solution of a specific location. Temporarily ignoring base station synchronization, the LTE signal structure itself (e.g., bandwidth and chip rate) supports a time difference accuracy that corresponds to about 20 meters (Sand, Mensing, and Dammann, 2007). The maximum allowed time difference between a pair of base stations, which is three µs, corresponds to a maximum distance of 900 meters, although a RMS value is surely much less. LTE is reported as being able to typically provide positioning accuracy of several tens of meters.[11] The best accuracy is attained for users within the boundary of the cell phone towers from which they compute their location. Users that lie outside the boundary could experience geometric dilution of precision that could make accuracy several times worse. Position accuracy implies time transfer with roughly 100 ns accuracy but in fact could be much worse, depending on the method by which each base station gets its own time. Differences from UTC could be ±1.5 µs according to ITU tolerances.

Where positioning is inferred to accuracies of tens of meters, the intrinsic time uncertainties in base stations will dominate the quality of time transfer to wireless users. That is, users should be able to maintain time to nearly ±1.5 µs of UTC as long as the base stations maintain time to this accuracy.

In the absence of GPS or other sources of UTC input, the accuracy of base station time with respect to UTC will degrade. Typical base stations may exceed ITU standards within one day. They are likely to maintain synchronization within tolerances for perhaps up to a year (Goode, 2018). Thus, although time quality would quickly degrade, positioning and navigation quality from LTE could remain of good quality for long periods of GPS outage.

It is worth noting that such benefits of LTE are likely to be available only to subscribers of commercial LTE providers. If one happens to be in a region served by other providers, cellular-based PNT might not be possible. However, at least one research group showed that it could exploit the data structure in cellular data without being a paying subscriber. They used an SDR to attain navigation accuracy that stayed on average within 12 meters of ground truth provided by a GPS track (Shamaei, Lhalife, and Kassas, 2016).

Implications for government and user investments

There may be few implications for any investments either by service providers or users in many cases, as long as users subscribe to an available service. Cellular-based PNT

---

[10] TDOA is used because this method does not require user equipment to have a precision clock for inferring range and because it does not involve two-way communications for measuring time delay. See later discussions on NTP and PTP.

[11] Reported in Koivisto et al. (2017). This might also include measurements of signal strength in the algorithms.

that does not require a subscription may require users to adopt new technologies within their cellular devices.

5G Cellular

Description and performance

Although the standards for 5G cellular telephony are not fully established, some of its characteristics can be anticipated. In particular, 5G will generally user higher frequencies and wider bandwidths than 4G and will attain efficiencies using multiple-input, multiple-output antennas to deliver higher data rates to many more users than can now be served by 4G. The higher frequencies will generally reduce distances over which data links can be established by large factors compared with today's cellular networks (Roessler, 2017). One consequence is the substantial densification of transmitter nodes to achieve the higher throughput and reuse of the spectrum by generating a large number of smaller cells (Koivisto et al., 2017). In addition there might be many indoor transmitters, such as in office buildings.

It is widely expected that the characteristics of 5G will allow users to determine their location within one meter or less. This is considerably better than now possible by GNSS alone. Moreover, there will be significant incentive for users and 5G providers to implement such capability.[12] Whether the computations to localize and track occur in user or provider equipment appears to be a choice that has not yet been made. There are implications for computational demands on user equipment and also issues for user privacy.

Much like 4G-LTE, the synchronization requirement for a 5G network is expected to be ±1.5 µs across the entire network, but the synchronization among adjacent nodes might be considerably better (Goode, 2018). Thus, while positioning accuracy could, in principle, support time transfer as good as a few nanoseconds, accuracy relative to UTC will depend on the method by which base stations get time. Those tolerances might not be too different from current 4G standards.

In a GPS or GNSS outage, the quality of synchronization will probably degrade only slowly. Thus, it is likely that positioning and navigation based on 5G can be sustained for long periods without GPS or GNSS, although the accuracy of time transfer with respect to UTC would probably degrade at the rate implied by the quality of any holdover clocks in the 5G network. For present discussion, we will assume that holdover performance is the same as 4G.

---

[12] As Koivisto et al. (2017) describe, one key use case of 5G is connected vehicles. From a user perspective, accurate location would support "smart traffic" management where traffic densities could be as much as 2,000 vehicles per square km. From the provider perspective, by accurately tracking vehicles, providers can provide desired data rates more efficiently; that is, with less power and/or bandwidth thanks to multiple-input, multiple-output beamforming techniques.

Implications for government and user investments

5G technology is coming. Over the next several years, the United States will probably see the first major deployments of 5G infrastructure, and there is no doubt that consumers will adopt 5G equipment over time. This technology could meet the PNT needs of many critical infrastructure sectors with few implications for acquisition other than those already expected under consumer market forces.

The only question may be how quickly the 5G deployments proceeds (Dignan, 2018) and the extent to which 5G will be available in remote areas where providers may not have an incentive to invest in the cost of dense transmitter networks.

Iridium

Description and Performance

Iridium has had a constellation of 72 communication satellites (66 operational and six spares in six orbital planes) at an LEO altitude of 780 km and a polar inclination of 86.4 degrees since 2002, with some early operational capability starting in 1998 (Graham, 2018). While this constellation has the primary mission of providing commercial mobile voice and data communications globally, it has recently been utilized to transmit PNT signals, thus providing a PNT solution from LEO that is independent of GPS. Iridium announced in May of 2016 that its STL system was ready for use as a PNT source that could serve as an alternative or backup for GPS (Reuters, 2016). STL is a subscription-based service of the partnership between Iridium and Satelles, a division of iKare Corporation with offices in Virginia and California. Satelles provides user equipment and service support to customers of the STL service while Iridium maintains the satellite constellation and its operations (Iridium, 2016). In December 2016, Satelles announced a further partnership with Orolia (the parent of companies Spectracom and Spectratime, which already are working in GNSS user equipment) to develop, market, and sell PNT solutions based on STL. Orolia has an existing GNSS user equipment solution, branded as Spectracom, and has now integrated STL as another source for PNT signals in that user equipment (GPS World Staff, 2017; Spectracom, 2018).

Since Iridium transmits in L-band between 1616 MHz and 1626.5 MHz (Jewell, 2016) it is not susceptible to jammers specifically targeting GPS signals, though a jammer could be made to cover those frequencies as well if it were regarded as important by an adversary. In addition, unlike GPS satellites in MEO where each GPS beam covers nearly an entire hemisphere of the earth (footprint radius of 7,900 km), each Iridium satellite has 48 narrow spot beams to cover a geographic area about one-third of that diameter (footprint radius of 2,500 km). The overlapping spot beams, in combination with randomized broadcasts, enable a unique method of providing location-based authentication of the Iridium signal that is very difficult to spoof (Satelles, 2016). Also, modern encryption used on the Iridium and STL signals—analogous to the GPS military-only signal, except that the STL signal is available to all commer-

Figure C.3
Iridium and GPS Orbit Comparison

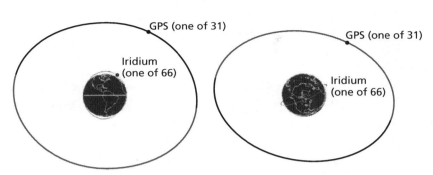

cial subscribers—adds to the difficulty of spoofing (Lawrence et al., 2017). Because the satellites are much closer to the earth, only about 4 percent as far away as a GPS satellite, and the broadcasts are focused in spot beams as stated above, the power of the signal at ground level is between 300 and 2,400 times (30–40 dB) stronger than GPS. Thus, even if Iridium-specific jammers were proliferated, they would need significantly more transmit power than a GPS jammer to be effective. Figure C.3 shows a diagram of an Iridium satellite orbit at 800 km altitude and a GPS satellite orbit at 20,200 km altitude.

STL receivers perform precise Doppler and range measurements at attenuations of up to 39 dB relative to unobstructed reception. Receivers can tolerate this much attenuation because the STL signal is modified from the standard Iridium transmission to enable more processing gain in the receiver.[13] Specifically, the quadrature phase-shift keying transmission scheme at the beginning of an STL "burst" transmission is designed to form a continuous wave marker that can be used by the receiver for burst detection and coarse measurement. Also, the subsequent data included in the burst is organized in pseudorandom sequences, which provides a way for receivers to carry out precise Doppler and range measurement by correlating the received sequence with the same sequence generated by the receiver. The data rate required by the receivers is reduced by having them carry the sequences onboard as well. This sequence correlation operation leads to a processing gain that enables the receivers to reliably operate in most deep indoor and urban canyon environments (Lawrence et al., 2017).

Another difference of the LEO Iridium satellites compared with the MEO GPS satellites is that the Iridium satellites travel at a much faster angular rate due to their shorter orbital periods (100 minutes versus about 12 hours) and much closer range. One advantage of this is that reflected "multipath" signals, which can be a significant

---

[13] The standard Iridium voice phone signal is not available indoors because it uses a two-way channel with lower system gain than the one-way channel broadcast for STL, and that two-way channel requires line of sight (Spectracom, 2018).

problem for GPS receivers especially in urban canyon environments, are less of a problem for STL receivers because the satellite geometry changes so rapidly that the multipath signals become more random, like white noise that can be more easily separated from the direct-path signal. Also, the faster angular motion of the Iridium satellites also leads to stronger Doppler signatures than GPS, which are useful for the positioning solution as well (Lawrence et al., 2017).

Yet another difference from GPS is that there are fewer LEO Iridium satellites visible from each location on earth due to their much lower altitude. While a user on the ground can have line of sight with 10–12 GPS satellites at a given moment, they can usually only see one or two Iridium satellites at a time. Hence, Iridium does not use the GNSS method of multilateration that requires a minimum of four visible satellites. While the details of the STL positioning algorithm are proprietary, it is publicly known that the system leverages Doppler ranging, as was previously used by the Transit polar constellation that was operational from 1964 to 1996 (Lawrence et al., 2017). That constellation required only four satellites in LEO because a position fix required observation of only one satellite at a time. As the LEO satellite moved overhead, its motion created a Doppler frequency shift in the signal received by a user on the surface (usually a ship at sea). The user's receiver had its own reference signal that it could compare with the received signal, allowing it to measure the frequency difference between the two. Since the measured frequency difference would change as the satellite moved overhead, the receiver could measure the change in the slant range vector (i.e., the Doppler shift projected on the line-of-sight vector to the user) over a given time interval (GPS World Staff, 2017). By combining several successive measurements with the data transmitted by Transit, which included the satellite's precise position as determined by ground control stations, the user could determine their own position. At that time, the computations could often take 30 minutes. STL combines the principle of Doppler ranging from LEO satellites with modern encryption and randomized spot beam broadcasts to improve on the performance of previous LEO-based PNT systems.

The STL positioning and navigation processing software updates an extended Kalman filter with a Doppler measurement and a range measurement. The range measurement provides position along the line of sight to the satellite (i.e., a circle around the satellite's nadir) and the Doppler measurement to further constrain a location. A single satellite pass can provide 3D position with better accuracy within the satellite's orbital plane than perpendicular to it. Improved accuracy is possible if more satellites are visible. Receivers on mobile platforms can also integrate measurements from IMUs for improved accuracy (Satelles, 2016).

The accuracy performance of the STL system is about ten times lower than GPS. Specifically, positioning accuracy is advertised as 30–50 meters (vertical and horizontal) and timing as 200 ns. For comparison, typical GPS performance provides about three-meters positioning accuracy and approximately 20 ns timing accuracy (Spectracom, 2018).

The primary advantage of STL is its ability to provide timing and position fixes when GPS signals are not available due either to shadowing, indoor attenuation, spoofing, or jamming. This positions STL as a complementary or backup solution to GPS, and Satelles and its partners have marketed receivers capable of interpreting both GNSS and STL signals.

The ability of receivers to acquire the STL signal indoors was tested in a high-rise building in 2017 (Lawrence et al., 2017). STL receivers were placed at multiple sites between the second and fourteenth floor. GPS receivers on upper floors near windows were usually able to track at most two GPS satellites, while those on lower floor could not track any. However, STL receivers experienced strong signals throughout the building, even on the lowest floor. The carrier-to-noise density ratio (C/N_0, a ratio of the carrier power to the noise power per unit bandwidth) for STL was between 35 and 55 dB-Hz, which is comparable to the GPS signal in an open-sky environment. They also tested indoor time transfer performance over a 30-day trial, measuring the difference in pulse-per-second outputs from an STL receiver and a GNSS "truth" reference with timing performance at least an order of magnitude better than STL. A typical timing errors was approximately 104 ns, with a maximum excursion of 471 ns. Lawrence et al. (2017) claim that the ability of STL to maintain sub-100 ns time keeping with a rubidium-based STL receiver has been demonstrated. They also conducted a 24-hour collection of position convergence measurements from an indoor STL receiver, finding that the receiver had an accuracy of better than 35 meters (vertical and horizontal) for 67 percent of the trials. Given sufficient time, an accuracy of 20 meters could be achieved in deep attenuation environments (Lawrence et al., 2017).

Because STL uses a separate satellite constellation from GPS, it represents a complete positioning backup, with the downside of reduced accuracy compared with GPS and the upside of stronger ground-level signal that is more difficult to jam or spoof and that may enable indoor access. GPS jammers would not affect STL receivers since STL utilizes a different portion of the spectrum. In principle, L-band jammers could be used to deny PNT solutions to STL receivers, although they would need to be considerably more powerful than the typical GPS jammer. Satelles has 25 ground stations around the world, each with a rubidium oscillator that is synchronized using GPS. These ground stations provide UTC to the satellites. In a GPS failure situation, the STL system would instead use those rubidium oscillators until GPS recovers, during which time the satellites would drift from the USNO master clock (Spectracom, 2018). Hence, STL does still have some timing dependence on GPS, though presumably it would be technically feasible to integrate an alternate master timing source into the system.

Implications for government and user investments

If STL were adopted as an alternative source of PNT, users would have to acquire user equipment. As an example, Orolia's Spectracom brand sells a "1204-3E: STL Input Module," which is a circuit board card that can be added to the existing Spectracom SecureSync GNSS time server and is marketed for increased time source resilience for critical services. The company also markets full GNSS PNT receivers with integrated STL receivers, such as the VersaPNT unit. Typically, these units are rack-mounted systems with a size comparable to a toaster (Orolia, undated). These devices are currently much bigger than smartphones, but since the STL frequency is so close to the GPS L1 frequency, we presume that user devices could be miniaturized even for such small devices given sufficient demand.

Users will also have to subscribe to the STL service to decrypt the signals. Iridium and Satelles claim that "many from industry and government are already using this service to achieve a more robust PNT solution," although information identifying those clients is not publicly available (Spectracom, 2018).

WWVB

Current Capability

Description and performance

The NIST provides several time services (Lombardi, 2003), one of which is a low-frequency broadcast with the radio station identifier WWVB. The WWVB broadcast is familiar as the signal used by radio-controlled clocks (RCCs) and watches that many consumers buy. Such RCCs are often marketed as "atomic clocks," although this is a misleading description.

WWVB transmitters are located in Fort Collins, Colorado (NIST, 2019).[14] They broadcast at 60 kHz. The pulse-width modulation of the signal requires only a few hertz of bandwidth, resulting in a data rate of one bit per second. The time message, which repeats every 60 seconds, is encoded at this data rate. The system uses two transmitters, each providing an effective radiated power of 70 kW, with a third standing by as a back-up.[15] WWVB is a source of UTC that achieves an accuracy of about 100 μs.[16]

A key issue for WWVB users is the high degree of EMI that can prevent signal reception (Lowe et al., 2011). A major source of interference is machinery and electronics, such as computers, but there are many other sources of man-made and natural EMI. Signal reception is typically much better at night than daytime, especially for

---

[14] WWVB should not be confused with NIST's WWV and WWVH signals, which are high-frequency broadcasts from Fort Collins and Hawaii, respectively. These signals provide less accurate time transfer than WWVB but can be received at much longer ranges.

[15] This power level is much increased from the original 1963 installation that provided only five kW of power.

[16] The range to a typical user would induce a time lag many times longer, but such time errors can, in principle, be partially corrected for user location (Lombardi, 2003).

users near the CONUS boundaries where the Fort Collins signal is weak.[17] To address these issues, NIST modified the signal in 2012 to improve user performance. NIST added binary phase modulation to the signal, which enables about ten dB detection advantage to receivers that exploit the phase information. Most legacy receivers will still work with the modified signal, although some legacy receivers that used a phase-locked loop to track the signal will no longer work. New receivers might also be able to accumulate received energy over time to further improve detection. An hour of accumulated energy would provide an addition factor of 60 (18 dB) of gain. Unfortunately, we are not aware of any studies that report how these amounts of gain would affect user reception by location or time of day.

Receivers that exploit the modified signal are now commercially available. The most common antenna used for these signals is a ferrite rod, which should be oriented favorably to maximize signal strength.

Implications for government and user investments

There are no implications for new national infrastructure to sustain the timing signal, unless there is a need to make this signal available to users in Alaska and Hawaii. There is already a market for RCC devices, and the improved signal might lead to an increase in the size of this market. Of course, many critical infrastructure users might need to acquire WWVB receivers.

Although WWVB antennas can be small enough for devices such as smartphones, the resulting performance might be inadequate, and the unpredictable orientation of antennas in such devices will be another source of signal loss. Thus, we think that WWVB might make the most sense for fixed (i.e., nonmobile) users that do not require a high degree of time accuracy that could use larger, fixed antennas.

Wired Time Signals

The majority of PNT alternatives above are one-way methods since receivers do not need to send back any information to transmitters regarding their own PNT solution. In this section on wired alternatives to GPS, all methods are two way. This is necessary because wired communications routes are often switched and routed and are subject to delays and latency of network traffic. Usually, the time to send a signal from one device to another is not even an approximate measurement of distance. Hence, all wired alternatives to GPS may provide time transfer, but they are not candidates for positioning and navigation.

---

[17] Other sources of EMI include distant lightning, multipath interference causing nulls, diurnally dependent atmospheric losses, shadowing and obstruction by geography, and on the East Coast, the U.K. "Rugby" signal from their similar timing broadcast. The orientation of receiver antennas can also be a source of loss.

All methods of wired time transfer are fundamentally based on the basic principle of NTP that is implemented on common computer operating systems (including Windows) to provide time from a server to its clients on a local area network. All attempts to measure the time delay for a signal to arrive from a master clock to a client computer is done by measuring four time markers (Klecka, 2015). The server sends a signal at time t_1 that is received by a client at t_2 (according to the client's clock, not the server's). The client sends back a signal at t_3 that arrives at t_4. The server computes the time to transfer a signal to the client as the average of the pair of time differences, $\delta = [(t_2 - t_1) + (t_4 - t_3)]/2$. Any systematic error in the client's clock (in values t_2 and t_3) cancels out, as do any other errors that are common to both paths. Finally, the server sends its time plus the correction δ to the client that now can compute a time that is synchronized with the server.

This method assumes that both paths are symmetric, and this is not strictly true in general. Errors in NTP time transfer arise from path asymmetries and time-stamping accuracy. The various methods described below essentially attempt to systematically improve on the quality of path symmetry and time stamping to achieve improved time transfer. PTP is a significant improvement on NTP, both of which are typically used in and across computer networks. However, PTP has also been applied to time transfer over large distances using the optical fiber of telecommunications networks. The highest fidelity example of this is the White Rabbit system.

PTP can also be used for time transfer in wireless networks (e.g., to synchronize transmitters in cellular base stations). At least some telecommunications providers do this already. Wired applications have the advantage of being entirely free of risks from RF interference and can be used in indoor and underground applications where GPS or other RF signals are unavailable.

Network
Network Time Protocol
Description and performance
Developed in the 1980s, NTP is commonly used to distribute time from a computer server to its clients on a local area network. By design, it minimizes the use of computer resources to potentially provide time to up to many thousands of clients. It is built into commonly used operating systems like Windows. A common application is to provide time to the integer second. In typical situations, it can provide time to an accuracy of only about one millisecond.

A recent effort to understand the fundamental limits of NTP achieved 10–20 microsecond accuracy by carefully eliminating time asymmetries (Novick and Lombardi, 2015). Errors arising from time stamping in each computer and latency of accessing computer memory account for the remaining errors. Such good performance cannot be expected in most computer networks.

A key concern for this study is the source of time for the server. The source of time could be GPS or an internet time server. NIST maintains a time server that implements NTP that can be accessed by common operating systems (NIST, 2018). However, in NTP applications, the accuracy of this source is probably not critical since NTP accuracy is so low anyway. The main purpose of NTP seems to be synchronization of local networks, which could make NTP applications relatively insensitive to GPS outages.

Implications for government and user investments

There are probably few, if any, implications for acquisition since most potential users of NTP will have computers with NTP already available. Users might simply have to configure their systems to use NTP as a timing alternative if NTP meets their needs.

Precision Time Protocol
Description and performance

PTP, also known as IEEE 1588, was initially introduced in 2002, and a revised standard was issued in 2008. It relies on the same basic algorithm as NTP, but its implementation is quite a bit more complex (Klecka, 2015). It supports networks of networks to minimize asymmetries in end-to-end time transfer across multiple network nodes and allows the best possible synchronization among peers in a network. It uses specialized hardware to improve the quality of time stamping.

Unlike NTP that simply defines a server and client clock, PTP defines several types of clocks: the Grandmaster that serves as "truth" for the network; ordinary clocks that only receive time; boundary clocks that can serve time to subdomains; and transparent clocks that pass time to subdomains with as little effect as possible (adding only a time correction for network delay) but that are otherwise not serving time. The addition of boundary clocks and transparent clocks make PTP scalable. It supports both end-to-end and peer-to-peer time transfer for tight synchronization within subdomains.

PTP achieves about one microsecond accuracy. It would meet the accuracy needs of the telecommunications industry. It is now used by at least some telecommunications users as well as a wide range of industrial, commercial, and scientific users. Like NTP, it could function almost entirely independent of GPS. If the Grandmaster clock is a high-fidelity source of UTC, such as a clock at NIST, then the network could sustain its one microsecond accuracy independent of GPS.

Implications for government and user investments

Any acquisition costs would largely be borne by users. In this context, users could be the end user or perhaps telecommunications providers that maintain and operate nodes in the backbone network.

PTP over Optical Fiber
Description and Performance

A recent study sponsored by DHS sought to test the limits of PTP over long distances between telecommunications nodes connected by optical fiber (Weiss et al., 2016). The key feature of this experiment was the ability to calibrate out the time asymmetry as there was a high-quality atomic clock at each end node: an atomic clock at NIST in Boulder, Colorado, and the USNO Alternate Master Clock for GPS at Schriever Air Force Base. The total distance was approximately 150 km. The total time delay was about double the delay implied by the speed of light in an optical medium, implying buffering in the telecommunications network. There was about 40 µs of path asymmetry, but this asymmetry remained fairly stable over the roughly 30-day experiment. With asymmetry removed, an overall timing stability better than 100 nanoseconds was observed. An experiment to assess the performance over a distance of 1,700 km (from Boulder, Colorado, to Chicago) is in progress (Cosart and Weiss, 2017).

It appears that this method can support such accurate time transfer over significant distances only if path asymmetry is well measured. While it is not plausible that this could be accomplished for all fixed users, it might be plausible for existing users with high-quality clocks that are traceable to UTC.

Implications for government and user investments

PTP over optical fiber seems especially appropriate as a method of backing up GPS time transfer for sites already using highly accurate clocks. For these users, the implications for acquisition are probably few, presuming that they already have optical fiber connectivity. The main resources required might be the time and effort to invest in a measurement campaign to quantify time asymmetry in the optical fiber connections.

White Rabbit
Description and Performance

White Rabbit technology development began at CERN in 2008 to provide subnanosecond time transfer for particle physics applications over distances of about 10 km, which is farther than the distance between nodes in a typical PTP subdomain (Serrano et al., 2013). This accuracy is significantly better than the few nanoseconds that is possible using GPS. This technology now has been demonstrated and commercialized. By design, it is intended as a high-accuracy extension of IEEE 1588, and it may soon be officially incorporated into the third release of the PTP protocol. It is backward compatible with earlier releases of IEEE 1588.

In White Rabbit, there are only two types of clocks: the master clock, which is the source of time for the entire network and switches, which are nodes in the network that can receive time from superior nodes and disseminate time to subordinate nodes. Nodes communicate using Ethernet protocol, and each switch has separate downlink and uplink connections. White Rabbit achieves much more accurate time transfer than earlier implementations of PTP by adding two key advances that improve time-stamping errors that are a key source of inaccuracy in PTP. First, unlike other PTP implementations in which boundary clocks are independent of the Grand Master Clock, here all clocks are syntonized (i.e., frequency disciplined) by a signal from the master clock that is communicated throughout the network.[18] This essentially eliminates any clock drift that could occur during forward and backward messaging in typical PTP implementations. Second, time stamping in the switches is based not only on the time message, but on the phase of the encoded signal. This improves the precision of time stamping and allows for a more accurate computation of time delay between nodes. The resulting accuracy of time transfer depends on how many levels of nodes are used, but even in the earliest demonstrations in 2013, White Rabbit consistently performed with better than 1 ns accuracy, often several times better, sometimes even achieving accuracy as good as 200 picoseconds. A recent demonstration (Kaur et al., 2017a) using upgraded White Rabbit hardware[19] over a 500-km link, constructed of four levels each 125 km apart, still exceeded GPS accuracy.

White Rabbit master units and switches are sold commercially by Seven Solutions, Inc. These units are rackable in standard 1U 19 inch racks. It is not clear whether Seven Solutions is yet selling the upgraded hardware that was used in the 500 km demonstration.

White Rabbit technology was developed by an international collaboration using open hardware and open software principles. The nonproprietary nature of this technology could facilitate evolutionary improvements or expansion into broader markets.

The master unit's source of time could be GPS, but also could be any other high-quality source like NIST's UTC atomic clock standard. If the network is independent of GPS, then its performance will be unaffected by any GPS outage.

Implications for government and user investments

White Rabbit would be most appropriate for use at sites that are key nodes for time distribution or users of very accurate, atomic clock quality timing that would benefit from even better accuracy. Less accurate methods, like ordinary PTP, could provide timing to the vast number of other users, like mobile and cellular users. A secondary benefit to such users might be the improved quality of PNT that is possible when high-quality timing is brought closer to the point of need.

Although we are not aware of White Rabbit being used over commercial fiber networks, it seems plausible commercial fiber could work as long as time delays due to network traffic, switches and buffering, could be avoided at key White Rabbit nodes. However, before White Rabbit was selected for widespread use, it would be important to verify that commercial fiber was a viable low-cost option. Any requirement to lease

---

[18] That is, all clocks will beat at the same frequency, even if they are not synchronized.

[19] The main upgrade seems to be switches that have three times the bandwidth for phase measurements.

"dark" fiber or to install dedicated fiber could involve costs that go beyond the White Rabbit technology itself.

User Equipment Alternatives to GPS

Holdover Clocks
Description and Performance
When GPS is degraded, holdover clocks can provide time when GPS goes out. Of course, integrating holdover clocks into user equipment is not new; virtually all user equipment has clocks capable of holdover for brief GPS outages. The accuracy of such clocks relative to UTC tends to degrade with time. Until recently, atomic clocks that could maintain less than one μs accuracy after 24 hours were the size of electronics boxes, costing at least a few thousand dollars and using at least tens of watts of power. The highest fidelity atomic clocks have performance orders of magnitude better, but the best performance is achieved only in laboratory-scale systems (Shkel, 2011; Kitching, undated).

In this subsection, we discuss the possibility of providing better holdover capability for supporting outages of longer duration. Beginning in the early 2000s, DARPA sought to develop an atomic clock capable of providing one μs accuracy after 24 hours in a device of one cm³ size, using less than 30 mW of power. The result was a chip commercialized by Symmetricom (now Microsemi) in 2011. This chip uses about 100 mW and has volume of about 17 cm³, and its accuracy appears to be close to the goal of one μs at 24 hours (Microsemi, 2018). The 2011 cost was about $1,500. The size and performance may be good enough for some critical infrastructure applications.

DARPA is continuing to develop other precision timing capabilities. In the Quantum-Assisted Sensing and Readout (QuASAR) program, DARPA is trying to make very accurate, laboratory-quality atomic clocks portable for use at field sites (DARPA, 2014). Such capability could potentially improve the holdover performance at important infrastructure sites requiring very accurate time, such as eLoran transmitters. We do not have information on the current status of this program. One goal of the QuASAR program is to transition the technology into another DARPA program called the Spatial, Temporal, and Orientation Information in Contested Environments (STOIC), one goal of which is to develop accurate atomic clocks for tactical military applications.

Implications for government and user investments
Acquiring and integrating holdover clocks is strictly a matter for user equipment. The DARPA programs will likely result in atomic clocks with a range of sizes and accuracies. Integration of such clocks into user equipment will have to be addressed on a case-by-case basis. However, it seems that the size and cost of even the smallest such clocks are still much too large for the most compact user devices, such as smartphones.

Inertial Navigation System

Description and Performance

An INS uses an IMU to measure angular and linear accelerations using gyroscopes and accelerometers. When these accelerations are integrated over time, the INS is able to compute changes in position. INS systems are essentially a holdover capability for position and navigation, assuming an initial calibration using GPS. INSs are routinely integrated into a wide range of devices, from smartphones to vehicles of all sorts.

As with holdover clocks, the DARPA Micro-PNT program has been pursuing the miniaturization of INS devices to provide tactically useful capability for GPS-denied environments. DARPA has specified "conservative" and "aggressive" goals for this development (Shkel, 2011). Even the aggressive goal results in location errors of about one nautical mile after an hour. While this level of performance is sufficient for many military applications (e.g., missile navigation lasting only a few minutes or flight of unmanned air vehicles aided by sensors). The current status of this program has not been reported.

Even the highest fidelity INS systems are not suitable backups to GPS by themselves. Position must be reinitialized at least every few minutes. In GPS-denied environments, SoOPs or other sensors might provide these position updates. For example, one study showed that a low-cost, commercial-grade INS aided by cellular CDMA signals could maintain navigation quality better than a tactical-grade INS alone (Morales, Roysdon, and Kassas, 2016). Indeed, it seems very likely that hybrid systems that combine INS with other data, such as maps and visual systems to recognize key landmarks such as street signs, can arrive at superior positioning solutions. A system based primarily of visual recognition of landmarks demonstrated effective navigation travel distances of ten km between updates (Chipka and Campbell, 2018).

We note one other potential use of improved INS systems: the detection of GPS spoofing attacks using differences between INS-based and GPS-based positions. However, INS systems are likely to be able to detect only the clumsiest of such attacks that suddenly introduce large position errors into GPS user equipment. More sophisticated attacks that slowly walk the GPS solution away from the true location probably would not be detected.

Thus, while INSs will continue to be needed in a wide range of navigation applications, improving their performance or reducing their size may offer limited value to most critical infrastructure applications.

Implications for Government and User Investments

The acquisition of INS devices is strictly a matter for user equipment. Developers of user equipment will undoubtedly take advantage of improvements that are cost-effective and useful. It is unlikely, however, that developers will implement such improvements as a response to the GPS threats envisioned in this study.

Atom Interferometer

Although atom interferometry was not proposed as a solution by any RFI respondents, we reviewed the academic literature on the potential for atom interferometers to provide drift-free inertial navigation by providing high-precision measurements of acceleration without the need for an external reference signal. In 2017, researchers at Sandia National Laboratories reported a new advancement that could allow atom interferometry to work at environmental temperatures, rather than at cold temperatures closer to absolute zero (Cartlidge, 2017). This new approach offers hope that atom interferometers might eventually work outside of laboratory environments.

The physics behind this technology is proven, but devices are still just beginning to be developed. DARPA projects using atom interferometers for position and timing. Multiple patent applications were granted in 2017 for atom interferometry, indicating that hardware and processing advancements are continuing.[20] However, we assess that these efforts are at technology readiness level 3.

Simultaneous Localization and Mapping
Description and Performance

SLAM is the solution to the problem of an autonomous vehicle moving through an unknown environment, in which the robot must build the map and estimate its position within it as it moves. This problem was formulated in the mid-1980s. Example applications are robots moving through unfamiliar (often indoor) environments (e.g., robotic vacuums) or rovers on Mars.

There are two major aspects to SLAM. The first is the selection of an algorithm that simultaneously estimates parameters of the map, the pose of the robot, and the position of the robot within the map. The second is the selection of sensors and signal processing that provides the data for estimation. Since the early 2000s, the algorithmic aspects of SLAM were considered to be fully solved (Durrant-Whyte and Bailey, 2006). Two algorithms, known as the extended Kalman filter (EKF-SLAM) and Rao-Blackwellized particle filters (FastSLAM), are commonly employed. These algorithms, in principle, solve the SLAM problem if the input data is correct. The key input data are the parameters describing "landmarks" that are observed by the sensors. As long as signal processing extracts landmarks from sensor data and correctly *associates* those landmarks on sequential views as well as when landmarks are revisited at later times, position estimates will remain as accurate as the sensor data supports. Accuracy can even improve over time. Some SLAM researchers have made SLAM algorithms available online.

---

[20] Three examples include: (1) Johnson et al. (2016); (2) Compton, Nelson, and Fertig (2018) ; and (3) A. Gill et al. (2018).

The most common types of sensor used in terrestrial[21] applications are visual or infrared imagery and lidar (Cadena et al., 2016). Imagery usually provides only angular measures to landmarks, and lidar provides range. These can be used independently or in combination. Software to extract landmarks from such data is also available online, although we have not assessed the generality or limitations of freely available software. Although SLAM is understood at the basic level, research continues, with the forefront being in making signal processing and SLAM algorithms efficient and accurate. These details appear to be very dependent on the nature of the application and environment.

A key emerging application will be the use of SLAM in autonomous vehicles. In such vehicles, millimeter-wave radar as well as visual imagery are likely to be key sensors. Autonomous vehicles will likely be a market force toward improving capability and cost reduction in many consumer applications. When GPS is available, SLAM will augment navigation performance by developing the local dynamic map, where the dynamic elements are other traffic or changes in the map, such as from construction.

It should be understood that SLAM algorithms are not needed if the map is fully known; it is needed only to the extent that there are dynamic elements in the map or if the autonomous vehicle goes off the map, as might be common for some emergency vehicles. When the map is known, position simply needs to be updated often enough that an INS can maintain the required accuracy.[22] We can ask: when GPS is absent, how much detail is good enough to have a map that is able to support street driving? Although we have not seen analysis, it seems to us that sensors will have to identify multiple landmarks on adjacent buildings and structures as vehicles move along streets, even in the absence of a dynamic environment. In an environment without GPS, autonomous vehicles will need to access data in a "navigation cloud" that includes enough information on landmarks to support the sensors that are continually imaging their surroundings. These landmarks would be adjacent buildings, structures, or terrain (or features on these).[23] Such "cloud-based-location-as-a-service" might be provided for free or by subscription from providers such as Apple or Google or as a government-owned service, akin to how NIST now supports timing services. It would

---

[21] SLAM has also been demonstrated in unmanned vehicles, both aerial and underwater. In the underwater case, acoustic sensors predominate.

[22] This case is probably better described as sensor-aided navigation, and it is the basis of alternative-PNT systems being developed for military applications. However, because of the potential for dynamic changes in many situations, SLAM algorithms will often be necessary. For simplicity, we will continue to use the term "SLAM" even in cases where sensor-aided navigation might be more precise.

[23] In autonomous vehicles, the "navigation cloud" would not really be essential when GPS is present. This is because GPS can provide adequate information about vehicle position on the street map. In this case, onboard sensors are used primarily to understand the dynamic environment in the immediate vicinity of the vehicle. The "navigation cloud" becomes essential for the vehicle to locate itself on a map and navigate long distances without GPS.

be natural for such a service to include data on the RF environment also, since location based on Wi-Fi and other signals are also useful for positioning and navigation.

In all likelihood, autonomous vehicles would access the "navigation cloud" using 5G connectivity. Without 5G, SLAM as an alternative for position and navigation would probably not be viable.[24] Note that 5G itself could potentially provide the time transfer support for a full PNT alternative. 5G might also provide positioning to complement SLAM-based estimates. Note, however, that unless 5G cellular capability has sufficient independence from GPS, SLAM could inherit GPS-related risks from connectivity requirements.

We cannot quote a typical accuracy for SLAM performance since the accuracy will surely be highly dependent on sensors and environment. It seems probable that systems that meet the requirements of a wide range of users could be developed. By design, they will meet the needs of those users. Certainly, SLAM for autonomous vehicles will be good enough to support safety requirements over short times and close distances, but the ability to sustain navigation over long periods without GPS will be contingent on the data and sensors that are available. Since we think that SLAM systems that are as good as GPS could be developed, at least in principle, we will later use five meters as a nominal performance value, recognizing that different performance values might be achieved for different user applications.

Implications for Government and User Investments

The major infrastructure requirement for SLAM to be an effective alternative to GPS is the establishment of a "navigation cloud" to host the sort of information that can be exploited by vehicle-based sensors. Whether this is a free service provided by companies like Apple and Google, fee-for-service, or government-owned, all are possible business models that could be explored.

Users also would have to acquire the sensors, computers, and algorithms necessary for SLAM navigation. Commercial market forces, such as the autonomous vehicle market, will probably suffice to mature SLAM further. Many vehicles, including human-operated vehicles, will probably have the sensors and computers to exploit SLAM for short-range vehicle safety. However, when GPS is absent, SLAM will only be able to support long-range navigation if the algorithms exploit supporting data, probably downloaded using 5G connectivity. Market forces are likely to lead to 5G-connected vehicles too, irrespective of its role in PNT solutions.

While the future availability of SLAM and 5G in the commercial market seems assured, it is less obvious that vehicles used in critical infrastructure sectors, such as manned emergency vehicles, will implement the same technologies unless SLAM was

---

[24] Although computers in user equipment probably could store imagery and for areas that vehicles frequently visit, it seems unlikely that affordable devices will be able to store all the supporting data for places vehicles might visit at the resolution required to provide lane-level navigation. With 5G connectivity, SLAM systems could download the required data as it is needed.

selected as a preferred alternative to GPS. For these users, this might imply an acquisition requirement.

Enhancements to GPS Resiliency

In this subsection, we address technologies that may enable the continued use of space-based GNSS signals, especially GPS, despite a threat to their PNT performance. These methods are not alternatives to GPS; however, like other alternatives, the result would be the availability of PNT at required performance levels. Since this report takes the position that operational consequences of threats and the cost of mitigations are more important than the particular means by which remediation is achieved, we here present the pros and cons of such "resiliency" options.

Below we discuss multi-GNSS chips and receivers, nulling antennas, and jammer direction finders. At least one other option, ready-to-launch spare satellites, belongs in this category. We do not analyze it because such an option is not actionable by DHS and because the complexity of this option goes far beyond the cost of the satellites themselves.[25]

Nulling Antennas
Description and performance
Antennas that suppress the energy of jamming signals, while admitting the desired GPS or GNSS signals with minimal attenuation, can be effective against noise jamming attacks. Such antennas are of three general types of increasing complexity and cost. Choke rings can reject jammers (as well as desirable GPS signals) near the horizon or below the horizon. These are likely to be inexpensive. Many of the advantageous effects of these devices might also be obtained from good antenna installation practices, such as by placing antennas high on buildings, but blocked from public view (DHS, 2017).

Nulling antennas, commonly called controlled radiation pattern antennas, are the next level of sophistication. These can substantially attenuate jamming signals. There are many such antennas sold commercially. The most complex of them can reject up to five or six simultaneous jammers in different directions, although many can only null one direction at a time. The most sophisticated antennas are beamforming antennas that maximize the signal in the desired directions but attenuate signals from all others, with the strongest attenuation in the jamming direction.

---

[25] Additional costs include launch vehicles and storage of satellites in clean environments as they increasingly age toward obsolescence. Complexities include the prompt mating of satellite spares with launch vehicles, transport, and rescheduling of launches that had already been planned for months or years in advance. Despite years of discussion and studies, the U.S. military has not yet figured out a cost-effective approach toward "operationally responsive space" as a backup to its critical, exquisite surveillance satellites.

Nulling antennas are typically at least several inches in diameter, too large for many personal devices like cell phones but suitable for vehicles or fixed installations. They are frequently used in military applications. Most are likely to do a good job of suppressing noise jamming. While many remain vulnerable to spoofing since such signals resemble authentic GPS signals, the most sophisticated of nulling antennas incorporate integrity monitoring and can successfully reject spoofing attacks (Jones, 2017).

The PNT performance that results when nulling antennas are effective should be comparable to performance in benign environments.

Implications for government and user investments

The implementation of nulling antennas as a mitigation to jamming threats is strictly a matter of user acquisition and integration. While it may be viable for important segments of users that employ GPS receivers as part of the fixed critical infrastructure such as telecom and the financial sectors, that drive or use larger vehicles like cars and trucks, and that use human portable systems such as those used by surveyors, these nulling antennas are too large and probably too costly for the use in the millions of handheld devices like phones, portable radios, and very small unmanned aerial vehicles.

Direction Finders

Description and performance

When jamming events occur, they can be detected and perhaps geolocated using direction-finding devices. This can be useful since in instances where statistics of GPS jamming have been collected, jamming events seem to occur at least several times per day on average (Curry, 2014). Since GPS jamming is illegal in the United States, this can allow follow-up actions by law enforcement when such events seem to be deliberate (e.g., the famous incident of the "white van man" who jammed GPS at the Newark, New Jersey, airport intermittently for many months in 2012) or remedial action against accidental jamming events (e.g., the accidental spoofing during a 2017 Institute of Navigation conference in Portland, Oregon). Such remedial actions can make jamming events less severe and allow holdover capabilities to sustain PNT during temporary disruptions. Without such remedial actions, jamming events might last too long for such holdover capabilities to remain effective.

There are GPS jamming direction finders sold commercially. They can be either handheld, such as was used to discover the jammer in the Portland, Oregon, conference center, or a set of networked antennas that can be deployed regionally and allows for near-real-time alerts (Curry, 2014). There have been several demonstrations of this latter capability, including a deployment at the 2014 Super Bowl (Rolli, 2015).

An important limitation of this mitigation method is that any enforcement action takes preparation, proper equipment for that function, and some time. The "white van man" was apprehended after about 12 months of intermittent jamming only after a significant surveillance effort (Curry, 2013). Longer periods of jamming and more significant assets dedicated to his capture would have probably yielded a better outcome. Even in a conference with hundreds of GPS experts and state-of-the-art equipment, the Portland spoofing event was not discovered and remediated for several hours[26] (Scott, 2017). Enforcement actions against deliberate and criminal jamming most likely would take longer than a few hours absent preparation for the event.

We conclude that direction finders, in and of themselves, may well be useful as a mitigation against long-term, high-power jamming, such as could be the case in a terrorist attempt to interfere with GPS. Unfortunately, they are likely to provide only limited value against jamming that occurs over brief periods, such as is characteristic of the common nuisance or negligent jamming. To be effective in this latter case, detection systems will have to operate in near-real time; other surveillance systems such as cameras that can help ID vehicles and individuals of interest and cell phone detection system that can ID phones such as cell phones present in the areas of interest could help eliminate the threat more rapidly than a simple jamming detection system. Handheld devices that only detect at short range are unlikely to be of much use except in sorting between a few candidate targets. Although networked antennas could be deployed regionally, they might not detect and geolocate the lowest power jammers, such as personal privacy devices, on scales as large as a city. Approaches that make use of virtual detection barriers that catch the emitter passing by or outposts to detected emitters within or near a perimeter could be useful.

The PNT performance that results when direction finders succeed in locating jammers will be as good as the benign environment once jammers are eliminated and will be as good as any holdover capabilities until then.

Implications for government and user investments

This case is unique among all technologies discussed in this report, insofar as acquisition of direction finders would be neither by providers of PNT signals nor by users. Rather, acquisition would be by law enforcement agencies. By themselves, direction finders add no value; their value attains only when jammer detections are followed by enforcement, with implications for manpower and other resources. Even if law enforcement chose to invest resources to combat GPS jamming, it is doubtful whether they could promptly mitigate specific instances of jamming by the most common sort of jamming event—the illegal use of personal privacy devices—since enforcement actions would likely take longer than the transient effects of jamming. Since this is more a matter of resourcing than of the capabilities of direction finders (albeit one aspect of evaluating this particular method of remediation), we assess direction finders as being of partial effectiveness (yellow) against threats of relatively short duration. We also note that over the long term, enforcement actions might have a deterrent effect on nuisance

---

[26] Though it could be argued that the very density of equipment made it challenging to sort out.

or negligent jammers that could eventually reduce the rate of jamming by personal privacy devices, even when remediation against individual events has limited success.

Summary

In this appendix, we examined a wide range of technologies for sustaining PNT performance. Across these technologies and options, the range of performance, the provision of full capability across PNT, and the relevance to different portions of the risk environment to GPS varies. Space systems respond to jamming and GPS specific risks, though all are exposed to the potentially high consequence risks of space warfare and intense space weather events. Ground-based wireless systems hedge more successfully against those sources of threat, though not all fully address ground-based jamming concerns and all are more accessible to threats. Wired timing signals all provide highly accurate time data and do so in a way that is largely not susceptible to the risks of concern to GPS (though with exposure to ground-based hazards). Finally, user-based equipment modifications can provide specific capabilities and address dependencies on GPS and do so in a way where adoption is driven by the individual needs (and willingness to pay) of end users.

Although many technologies show potential to serve as a backup PNT method for at least some risk cases for some users, not all technologies are equal. They differ not only in terms of available accuracy but also in other ways. For example, most will be available only in developed areas (e.g., with communications infrastructure), but a few could serve even remote areas. Some of the technologies can serve indoor users because the RF signals penetrate into buildings. This is due partly to the use of frequencies lower than GPS and partly to high signal strengths.

It is important to note that not all options are equally mature, however. Some are already on a path toward commercialization with an existing market demand, and others would probably be matured to maximum cost-effectiveness only if demand were driven via government action or subsidy. Even if market efficiencies are assumed, cost-effectiveness is likely to vary considerably.

Traditional Acquisition Versus Public-Private Partnership

At least one potential provider of an alternative PNT technology has suggested that a PPP might be preferable to a traditional government procurement approach to acquiring a GPS backup system. In general, a PPP approach involves the contractor in financing all or part of the investment cost of the project in anticipation of future revenue streams from the government and possibly other customers, whereas in a traditional procurement approach, the government pays the capital costs of the project. Depending on the type of PPP, the contractor may continue to own, operate, and maintain the project over the lifetime of the contract, or ownership, operation, and/or maintenance may revert to the government either immediately or sometime in the future.

In the United States and other countries, PPPs have been used to finance infrastructure projects such as toll roads, bridges, tunnels, and dams as well as other types of assets, including prisons, hospitals, and schools. The DoD has also privatized assets such as utility distribution systems, military housing, and visitor lodging to gain access to private-sector capital to upgrade neglected infrastructure and to private-sector management skills in activities that are not core defense capabilities.

Since the private-sector's cost of capital is typically higher than the government's,[1] PPPs should result in transfers of risk and efficiency gains that outweigh the higher financing costs to be cost-effective to the government. Under a PPP contract that requires the contractor to design, finance, build, operate, and maintain an asset in exchange for fixed annual payments after the asset becomes operational, the contractor will have incentives to

- construct assets within schedule and budget, since it bears the risk of construction cost overruns and delays
- minimize life-cycle costs as part of the design of the asset

---

[1] For example, OMB Circular A-94, Appendix C, indicates that the nominal interest rate on ten-year Treasury notes and bonds for calendar year 2018 is 1.8 percent (OMB, 2017), whereas the ten-year high quality market corporate bond spot rate for April 2018 is 3.96 percent (U.S. Department of the Treasury, 2018).

- maintain the asset to meet contractual performance requirements (U.K. National Audit Office, 2018).[2]

In the case of PNT technology, the contractor may also have access to additional revenue streams from customers that are not available to a government owner—for example, if the contractor can sell higher-quality or value-added services that are not available as part of a government-provided service to the general public. These other sources of revenue could reduce the investment costs that would need to be repaid by the government.

Government organizations typically use value for money (VfM) analysis to compare PPPs with traditional procurement approaches (see, for example, Kweun, Wheeler, and Gifford, 2018). VfM analysis is primarily a financial analysis from the perspective of the government agency that assumes the project has net social benefits and considers how it should be procured. The Public Sector Comparator (PSC) is a baseline scenario for the life-cycle risk-adjusted cost of delivering a project if the government finances, owns, and implements the project at its own risk. It can be compared with either a hypothetical PPP bid, also called a "shadow bid," or an actual bid from a PPP contractor. The projects should be compared using the same assumptions for timing, funding, procurement costs, output specifications, and performance standards. The cost components are shown in Figure D.1.

The PSC cost components include the net present value of the following.

- The base cost, which consists of direct costs, such as capital, operating and maintenance costs, and indirect costs, such as overhead and procurement costs, net of any revenue from the project.
- Retained risk is the value of any risk that is not transferable to the bidder under a PPP. It is the same for the traditional procurement scenario and the PPP scenario.
- Transferable risk is the value of any risk that is transferable to the bidder under a PPP but that would be retained by the government under a conventional procurement. It can be estimated based on an expected value or a Monte Carlo simulation using the magnitudes and the probabilities of the risks.
- Competitive neutrality refers to the value of any competitive advantages or disadvantages of the government agency relative to a private company that are not already captured in the other categories. These factors can include a lower cost of capital, access to tax-exempt debt, and foregone real estate, sales, or corporate taxes that would have been paid by the PPP contractor.

The PSC costs are compared with the sum of the net present value of retained risks and the annual payments to the contractor under the PPP scenario. The govern-

---

[2] Budgeted government organizations sometimes neglect infrastructure maintenance, which can result in higher future maintenance costs or shortened asset lifespans.

Figure D.1
Comparison of Public Sector Comparator and Public-Private Partnership in Value for Money Analysis

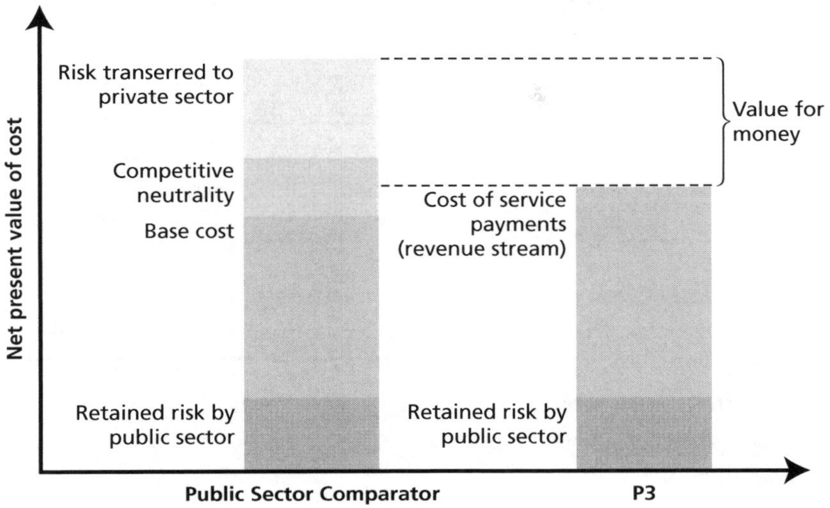

SOURCE: Jeong Yun Kweun, Porter K. Wheeler, and Jonathan L. Gifford, "Evaluating Highway Public-Private Partnerships: Evidence from U.S. Value for Money Studies," *Transport Policy*, Vol. 62, 2018, pp. 12–20; used with permission.

ment's procurement and oversight costs should also be added to the shadow bid or actual PPP bid for a fair comparison with the PSC. The option with the lower net present value of total costs would be chosen under the VfM approach.

DeCorla-Souza et al. (2013) argue that benefit-cost analysis (BCA) is a more comprehensive tool than VfM because it incorporates nonfinancial impacts such as benefits to users or nonusers of a project that may accrue from earlier delivery. A PPP approach may be able to deliver a project faster because it reduces delays in construction or financing due to incremental funding or limitations on government debt. In the extreme, a PPP may be the only vehicle for delivering a project if the government agency has a limited capital budget. A BCA also excludes costs such as taxes and tolls that are transfer payments, not economic resource costs, except to the extent that they change economic behavior.

Osei-Kyei and Chan (2015) reviewed the literature[3] on critical success factors for PPPs from 1990 through 2013, looking across countries, sectors (such as water, transportation, telecommunications, and energy), and the stage of the project (feasibility, initial design, and business case analysis). The factor mentioned most frequently was

---

[3] Their literature search focused on nine journals in the construction, project management, and engineering fields, and they conducted a content analysis of 27 publications on critical success factors for PPPs. Research approaches included case studies (41 percent), surveys or questionnaires (37 percent), and mixed methods (a combination of interviews, case studies, and/or questionnaires; 22 percent).

appropriate risk allocation and sharing between the government and the contractor. During negotiations, project risks should be clearly defined and allocated to the party that can best mitigate them. The second most frequently mentioned success factor was a reliable and well-structured contractor or joint venture with strong technical, operational, and managerial capacity to execute the project. Other frequently mentioned success factors were long-term political support for a PPP approach (so that contracts would not be undercut by a change in government or administration); public or community support (including the media, trade unions, and other nongovernmental organizations as well as the general public); and a transparent procurement process, including frequent communication between the government, contractor, and external stakeholders.

There are relatively few evaluations of the benefits of completed PPP projects. McNichol (2013) cites a 2007 study by Infrastructure Partnerships Australia that examined PPP projects constructed from 2000 to 2007. It found that on average, 18 percent of traditionally procured projects had schedule overruns in comparison with 10 percent of PPP projects. With respect to costs, 45 percent of traditionally procured projects had cost overruns compared with 14 percent of PPP projects. For projects that had either schedule or cost overruns, on average PPPs had smaller overruns than traditionally procured projects. The U.K National Audit Office (2018) reports a lack of data available on the benefits of PPP, although surveys of project managers and government departments indicated that private finance initiative (PFI) projects were delivered within budget more often than non-PFI projects and that greater certainty about construction costs was a benefit of PFI. Operating costs were reportedly similar or higher for PFI projects, although it is difficult to control for factors such as risk transfer and differences in standards between contracts. Contractually agreed standards under PFI have resulted in higher maintenance spending, which is seen as a benefit because government organizations tend to underfund infrastructure maintenance. However, the National Audit Office noted that some of these benefits could have been achieved under traditional procurement, using fixed-price construction contracts or long-term maintenance contracts, for example.

In the short term, PPPs increase government organizations' budget flexibility and spending power because no up-front capital outlay is required. However, they also create long-term financial commitments that can limit future budget flexibility, particularly if the number of PPP contracts increases over time. The National Audit Office found that long-term contracts were sometimes costly to modify if additional capital investment was needed, and in some cases government agencies were locked into paying for services that were no longer required. In addition, contract termination can be expensive, particularly if the financial structure of the PPP venture requires additional fees for early repayment of the debt. To address some of these issues, the U.K. government revised its guidance on PPPs in December 2012 and renamed the program Private Finance 2 (PF2).

In general, quite specific data on the costs and risks associated with procurement of a system would be needed to determine whether a PPP would be more cost-effective than traditional procurement. If the federal government determines that any system would be socially beneficial, it could compare alternative procurement mechanisms using VfM analysis based on a hypothetical shadow bid or actual bids submitted by contractors for that specific system. Alternatively, the implications of different funding mechanisms could be incorporated into an overall BCA framework to address which approach would deliver greater estimated net benefits to society. Such analysis naturally would occur after these federal decisions and, in some cases, after the marketplace has identified a commercially successful system.

For our case of an alternative or complement to GPS, we would expect an obvious, economically dominant opportunity for a PPP. The government could seek to expand the scale of systems that are otherwise being deployed in limited areas. It appears likely that some system—5G or MBS or something else entirely—will be deployed in large urban areas, providing more precise position locations of and for cell phone users. The federal government could simply allow the marketplace to determine the winning solution and then leverage that solution into a larger-scale deployment if they so desire. In this case, it would be most natural for the provider of the commercial system to maintain and operate any expansion, allowing economies of scale and ensuring interoperability as the commercial system are upgraded.

Initial deployments of these new systems are naturally being targeted to the areas where they offer the highest economic payoff. These areas tend to be higher-income areas because this is where the ads being sent to the cell phone users are more valued.

In particular, and as discussed in the main body of this report, a system similar to NextNav MBS is likely to be deployed in many urban areas. Another federal PPP, FirstNet, is required to provide floor-level information for first responders, something which MBS can provide. Expanding the NextNav deployment, or whatever system FirstNet chooses, would cover more areas with a backup PNT system—one which promises both to be easily integrated into cell phones, a requirement for FirstNet that also promises wide accessibility, that also offers improved performance such as operating inside buildings and modestly improved accuracy. Moreover, the planned deployment for FirstNet should cover a large fraction of the U.S. economy, thereby *already* providing a backup for that much economic activity. Additionally, the established FirstNet.gov organization allows an institutionally easy path for expanding the PPP.

Such a PPP would expand the economic benefits from a higher accuracy system to more people and also enhance some governmental functions, such as emergency services in additional areas. But this choice must await either the decision by FirstNet or the marketplace determination of a successful system.

U.S. Commitments and Obligation to Sustain and Operate GPS

The federal government has made strong commitments for the sustainment of GPS and the maintenance of backup capabilities through various statutes and policies and by seeking public and industry comments through the Federal Register and by implementing various performance standards.[1] Title 10 of the U.S. Code, Section 2281, tasks the Secretary of Defense to sustain and operate GPS services for the national security interests of the United States as well as for peaceful, civil, commercial, and scientific uses on a continuous worldwide basis free of direct user fees. This statute requires that such sustainment and operation of the civilian GPS system known as the Standard Positioning Service (SPS) will meet the performance requirements of the Federal Radionavigation Plan. In addition, Title 51 of the U.S. Code, Section 50112, conveys a sense of Congress in which Congress encourages the President to ensure the operation of GPS on a continuous worldwide basis free of direct user fees and provide adequate resources to achieve and sustain efficient management of the electromagnetic spectrum and protect the spectrum from disruption and interference. These two statutes demonstrate that the federal government has very strong intentions to sustain GPS. Additional statutes provide less specific support but authorize broad mission requirements to agency heads, which may involve PNT services. These requirements include but are not limited to air and marine navigation, critical infrastructure resilience and security, and space transportation systems, among other things.

Although statutes commit the federal government to sustain GPS, most of the federal government's intent to sustain and operate GPS is codified and reinforced in various policies and executive orders. Certain policies also recognize the need for backup capabilities to GPS. Table E.1 includes relevant examples.

While not specific requirements or commitments, GPS performance standards also provide goals to help ensure the sustainment and operation of GPS. These metrics demonstrate desired goals related to operational satellite counts, position service availability, and position accuracy, among other areas. According to Renfro, Terry, and

---

[1] Such federal commitments cannot guarantee future appropriations, even to sustain GPS, and as such, can only indicate the current intention of the government.

Table E.1
GPS Commitments in Selected Federal Policies

Policy	Examples of GPS and Backup Capacity Commitments
2017 Federal Radionavigation Plan	• Reiterates intent to sustainment and operation of GPS • Notes recognition of "the benefits of providing a backup capability to GPS to mitigate the safety, security, or economic effects of a disruption of GPS service"
2010 National Space Policy	• Reiterates the GPS requirement • Requires that the United States operate and maintain GPS consistent with published performance standards and interface specifications and to implement, as necessary and appropriate, redundant and backup systems or approaches
2004 U.S. Space-Based Positioning, Navigation, and Timing Policy/National Security Presidential Directive-39	• Includes the GPS requirement and states that the fundamental goal is to ensure PNT services, augmentation, backup, and service denial capabilities that provide uninterrupted availability, among other things • Contains associated requirements for the Secretaries of Transportation and Homeland Security
1996 GPS National Policy	• States the intent "to provide GPS SPS for peaceful, civil, commercial, and scientific use on a continuous worldwide basis, free of direct user fees"

Boeker (2017), only one metric related to the notification time for scheduled interruption was not met in 2016. Specifically, the metric that asserts at least 48 hours' notice will be provided for scheduled interruptions was met in 27 of 28 cases. In the case of the exception, only 45 hours' notice was provided. Additional performance levels exist for users of Wide Area Augmentation System and Precise Positioning Service. The GPS Civil Monitoring Performance Specification (DoT, 2009) provides requirements for monitoring the GPS civil service and signals based on top level requirements to monitor all signals all the time. The federal government also seeks to place satellites in primary orbital slots so that a minimum of six satellites will be in view of users anywhere in the world at any time.

Solar Storm Effects on the GPS Satellites

This appendix expands on the earlier discussion in Chapter Three of the potential impact of a large solar flare on the functioning of GPS. In particular, it focuses on how the GPS satellites themselves would experience both the prompt and lingering effects of a massive solar storm. A flare event generates a combination of direct radiation that propagates at the speed of light, some particles that move at relativistic speeds, and particles at substantially lower speed that enhance the natural radiation belt the GPS satellites operate within. In general, we find the GPS satellites themselves are unlikely to be significantly affected by even the largest solar events.

A solar flare, an explosion at the surface of the sun (the photosphere), generates a burst of electromagnetic radiation, including intense radio waves, X-rays, and gamma rays, which travel outwards at the speed of light, reaching earth in about eight minutes. As noted in Chapter Three, these radio bursts can heat the earth's upper atmosphere, increase ionization, and generally lead to an enhancement of electrical currents and electron density in the ionosphere. This changes how satellite radio transmissions refract through the atmosphere, especially disrupting high-frequency communications signals. For GPS, this "scintillation" in the upper atmosphere is a common cause of "loss of lock," where ground-based GPS user equipment loses reception, making GPS ineffective for periods of tens of minutes up to about three days, depending on the duration and recurrence frequency of the solar events. However, there are no significant direct effects on the satellites themselves from this radiation.

The flare event can also generate SEPs, particles with energy greater than 1 megaelectronvolt (MeV). These particles, traveling at a range of relativistic speeds, can reach the earth system in tens of minutes to a few hours after the flare, contributing to increased total ionization dosage of satellites, increasing the rate of "single event upsets," and the aforementioned heating and ionization of the ionosphere that disrupts radio communications.

The bulk of solar plasma material expelled from the sun in a CME (which is sometimes coincident with a solar flare) takes significantly longer to travel through space and is not always aligned with the earth's azimuthal position in its orbit around the sun. When it is aligned, it can reach earth within 14 hours to four days, depending on the variable speed of material from the event. The effect at the earth depends on the CME speed and the orientation of the interplanetary magnetic field in the CME, which can help or hinder the injection of material into the magnetosphere through magnetic reconnection. Typical results of major CME-driven geomagnetic storms are particle precipitation at the magnetic cusps (visible as aurora), injection of high energy particles into the ring current and radiation belts, and increased radiation, total ionization dosage, and probability of single event effects for satellites depending on orbital altitude (Royal Academy of Engineering, 2013).

Chapter Three describes the Carrington event of 1859, apparently the largest flare and CME in the historical record. Below, we use that as a limiting case for assessing the potential effects of a solar storm on the GPS satellites.

Satellites in the space environment are always subject to ionizing radiation in the form of electromagnetic waves (X-rays and gamma rays) as well as high energy particles, which can degrade photovoltaic panel efficiency, embed gradually increasing charge in insulating material, eventually leading to electrostatic discharge and even impact logic boards, creating "single event effects" where a circuit may be overloaded or a bit may be flipped by a single high energy particle impact, causing a range of potentially debilitating damage to the satellite.

There are several strategies employed by satellite designers and operators to mitigate the effects of the radiation environment in which their satellites fly. One is redundancy, where satellites are designed with multiple backup or error correction systems. One example would be the triple modular redundancy system used for the flight computers on the U.S. Space Shuttle, where three processors performed calculations in parallel and compared answers to check for single event upsets in one of the processors (Siceloff, 2010). Another mitigation strategy is "safing," where operations of the satellite are suspended when the spacecraft is passing through a region or time of expected inundation by higher energy particles in an effort to lower the electrical load of onboard circuitry or turn off particularly vulnerable sensors and make single event effects less likely. This is rarely used on satellites providing a commercial or government service as it effectively guarantees a service disruption, while the decision not to "safe" the satellite carries the chance that the satellite will continue to operate unaffected. One example is the suspension of observations by the Hubble Space Telescope when it passes through the South Atlantic Anomaly (SAA), a region of enhanced energetic particles in LEO. This is done because the fine guidance sensor cameras on Hubble could be overwhelmed by high energy particles in the SAA (Gonzaga, 2012).

The most effective and commonly used mitigation strategy is the design practice of encasing sensitive satellite components (especially solid-state electronics) in additional metal (usually aluminum) to absorb ionizing radiation and high energy particles. This is known as "shielding," and is typically expressed in terms of the thickness of aluminum used. For example, 200 "mils" is 200 milli-inches thickness of aluminum or 0.508 cm. Since aluminum has a density of 2.71 g/cm^3, one can also express the thickness as an area density of 2.71 g/cm^3 × 0.508 cm = 1.38 g/cm^2.

Generally, satellites are designed to withstand the radiation environment in which they will be flying, and this drives the decision of how much shielding to apply. For example, when considering concepts for a spacecraft to be sent to orbit Jupiter's moon Europa, NASA scientists estimated the radiation environment at Europa to be "~2.9 MRad total ionizing dose (TID) behind 100-mil-thick aluminum," also noting that "this level is more than twice that required for U.S. defense systems that are required to operate through nuclear explosions, and seven times greater than any previous NASA mission" (Keller, 2011). And so spacecraft designers ought to use electronics that can operate under that assumed TID at that shielding level, or else add more shielding.

Engineers designing satellites for operation in earth orbit consult empirical and theoretical models for the TID over a given period of time (e.g., per year) at different orbital altitudes and inclinations. Table F.1 shows values for the TID from natural environment expected for a circular orbit with a 30-degree inclination and a GPS-like altitude of 20,200 km. The actual hardening levels used in GPS satellites are not known at the unclassified level and therefore not used here. However, one can infer that GPS satellites would need to be shielded to withstand the radiation environment in which they fly, which is a rather intense one in the outer electron radiation belt.

We note that, assuming 5 mm ("200 mil," or 1.38 g/cm²) thickness of aluminum shielding, the estimated TID from one year in a 30-degree medium earth orbit (MEO) is about 8×10^4 rads per year. This is an average of 2.2×10^2 rads per day (Wertz and Larson, 1999).

Jiggens et al. (2014) modeled estimated TID for the Carrington event as a function of aluminum shielding thickness (see Figure F.1). For that same 5 mm thickness of aluminum shielding, they estimated a TID of about 1.5×10^3 rads over the one to two days of the Carrington event geomagnetic storm. That is a range of 7.5×10^2 rads per day to about 1.5×10^3 rads per day, roughly three to seven times the average daily dosage according to the natural source model shown in Table F.1.

Table F.1
Total Ionizing Radiation Dosage for Satellites in Medium Earth Orbit (30 degree inclination, altitude 20,200 km)

Aluminum Shielding Thickness		One Year TID
(mils)	(mm)	(Rads)
100	2.54	1,000,000
200	5.08	8,000
300	7.62	700

SOURCE: James R. Wertz and Wiley J. Larson, *Space Mission Analysis and Design*, 3rd ed., Torrance, Calif.: Microcosm Press; Dordrecht, Netherlands: Kluwer Academic, 1999.

Figure F.1
Modeled Carrington-Type Event Could Result in Total Ionizing Dosage of ~1.5 × 10³ Rads (Si) Over 1–2 Days or About 3 to 7 Times the Normal Daily Dosage for a Medium Earth Orbit

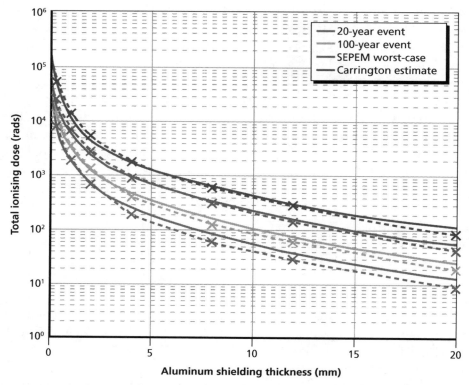

SOURCE: Jiggens et al., "The Magnitude and Effects of Extreme Solar Particle Events," *Journal of Space Weather and Space Climate*, Vol. 4, 2014, Figure 10 (CC BY 4.0).

While significantly higher than the average daily dosage at MEO, the estimated TID for a Carrington event is still less than 2 percent of the annual TID. Given we expect that GPS satellites are hardened to withstand the natural TID for their orbit in MEO, it seems unlikely that the TID from a Carrington-level event alone would pose a significant threat to the health or estimated lifetime of the GPS satellites themselves.

In fact, there is reason to expect that GPS satellites are hardened to higher levels than the equivalent of 5 mm aluminum. As mentioned earlier, defense satellites like GPS are typically designed to survive the enhanced radiation dosage that would occur in the earth's radiation belts if a nuclear weapon were detonated at high altitude. Such a detonation has been estimated to deliver approximately 8×10^5 rads over 30 days at 20,200 km altitude, behind 2.5 mm of aluminum shielding, an average of about 2.7×10^4 rads per day (see Wertz and Larson, 1999, Fig. 8-15b). This value would likely be significantly lower behind 5 mm of aluminum shielding instead of 2.5 mm, but even if it were a factor of 10 lower for an average of about 2.7×10^3 rads per day, that is still significantly more intense than the estimated TID from a Carrington-level event. Hastings and Garrett (1996) estimate a "hardened DoD spacecraft" to experience a TID of 1.5×10^5 rads behind 3 g/cm^2 of shielding (over 10 mm thickness of aluminum) for a "saturated nuclear environment after one year."

While admitting that a range of estimates exist for these theoretical events, it is clear that, for the same shielding levels, the TID threat to GPS satellites from a high-altitude nuclear detonation seems much more significant than from a Carrington event and will likely last for a longer period of time. If GPS satellites are indeed hardened to withstand such nuclear-enhanced radiation belt dosage, they ought to be easily able to withstand the TID from a Carrington-level geomagnetic superstorm as well.[1] And as noted above, their hardening to the ambient space environment that they certainly experience appears more than adequate to protect the GPS satellites from even a Carrington-scale event.

The primary GPS-relevant risk of a Carrington-level event would thus be an up-to-three-days outage of GPS availability due to modification of the ionosphere by radio bursts, X-rays, and SEPs, which is the focus of Chapter Three.

---

[1] We note that the described discussion focuses solely on gradually accumulated TID. However, a Carrington-level event (or a nuclear detonation) also increases the probability of single-event effects at a given satellite, which are difficult to incorporate into predictive models.

Bibliography

Abt Associates, *Social and Economic Impacts of Space Weather in the United States*, Bethesda, Md.: Abt Associates, September 2017.

ACIL Allen Consulting, *Precise Positioning Services in the Construction Sector*, Melbourne, Australia: ACIL Allen Consulting, June 2013. As of October 13, 2019: http://www.ignss.org/LinkClick.aspx?fileticket=ntyClJz4fh8%3D&tabid=56

Adamchuk, Viacheslav I., Mark L. Bernards, George E. Meyer, and Jerry A. Mulliken, "Weed Targeting Herbicide Management," *Precision Agriculture* (University of Nebraska Lincoln Extension), EC708, 2008. As of June 26, 2016: http://extensionpublications.unl.edu/assets/pdf/ec708.pdf

Agriculture.com Staff, "Things to Know Before You Buy a Guidance System," *Successful Farming*, July 6, 2010. As of June 26, 2018: https://www.agriculture.com/machinery/precision-agriculture/gps-guidance/Things-to-know-before -you-buy-a-guidance-system_236-ar6436

Alho, Timo, Michael Hickson, Tero Kokko, and Tommi Pettersson, *Conversion to Automated Straddle Carrier Terminal*, Helsinki, Finland: Kalmar, 2015.

Alliance for Telecommunications Industry Solutions, *GPS Vulnerability*, Washington, D.C.: ATIS, ATIS-0900005, 2017. As of October 13, 2019: https://access.atis.org/apps/group_public/download.php/36304/ATIS-0900005.pdf

ATIS—*See* Alliance for Telecommunications Industry Solutions.

Baker, Jim, "Positive Train Control," Littleton, Colo.: Joint Council on Transit Wireless Communications, White Paper, May 2012.

Balduzzi, Marco, Alessandro Pasta, and Kyle Wilhoit, "A Security Evaluation of AIS Automated Identification System," *ACSAC '14: Proceedings of the 30th Annual Computer Security Applications Conference*, New Orleans, La.: Association for Computing Machinery, 2014.

BEA—*See* Bureau of Economic Analysis.

Betters, Elyse, "Apple's iBeacons Explained: What It Is and Why It Matters," Pocket-lint, September 18, 2013. As of October 13, 2019: https://www.pocket-lint.com/phones/news/apple/123730-apple-s-ibeacons-explained-what-it-is -and-why-it-matters

Bhatti, Jahshan, and Todd E. Humphreys, "Hostile Control of Ship via False GPS Signal: Demonstration and Detection," *Navigation*, Vol. 64, No. 1, Spring 2017.

Blackwell, Thomas H., and Jay S. Kaufman, "Response Time Effectiveness: Comparison of Response Time and Survival in an Urban Emergency Medical Services System," *Academic Emergency Medicine*, Vol. 9, No. 4, April 2002.

Blazyk, Janet M., Chris G. Bartone, Frank Alder, and Mitchell J. Narins, "The Loran Propagation Model: Development, Analysis, Test, and Validation," London: Royal Institute of Navigation, NAV08/ILA37, 2008.

Board of Governors of the Federal Reserve System, "Foreign Exchange Rates—H.10," webpage, 2019. As of July 2, 2019:
https://www.federalreserve.gov/releases/h10/hist/default.htm

Breslin, Sean, "Delta Lost $125 Million in Five Days Following Recent Atlanta Storms," The Weather Channel, April 17, 2017.

Broadcom, "BCM47755: Third-Generation GNSS Location Hub with Dual Frequency Support," webpage, undated. As of October 13, 2019:
https://www.broadcom.com/products/wireless/gnss-gps-socs/bcm47755

Brunker, Mike, "GPS Under Attack as Crooks, Rogue Workers Wage Electronic War," NBC News, August 8, 2016. As of June 26, 2018:
https://www.nbcnews.com/news/us-news/gps-under-attack-crooks-rogue-workers-wage
-electronic-war-n618761

Bureau of Economic Analysis, "RIMS II Multipliers (2010/2010): Table 2.5," in Regional Input-Output Modeling System (RIMS II), Washington, D.C.: BEA, 2010. As of October 13, 2019:
https://www.nrc.gov/docs/ML1326/ML13262A071.pdf

———, "Industry Data: GDP-by-Industry," webpage, 2018a. As of October 13, 2019:
https://www.bea.gov/industry/gdpbyind_data.htm

———, "Regional Economic Accounts: Downloads" (under GDP: Annual GDP by Metropolitan Statistical Area), dataset, June 11, 2018b. As of June 11, 2018:
https://www.bea.gov/regional/downloadzip.cfm

———, "Gross Domestic Product," *National Economic Accounts*, 2019. As of October 13, 2019:
https://www.bea.gov/national/index.htm#gdp

Bureau of Transportation Statistics, "Miles of Freight Railroad Operated by Class of Railroad," webpage, July 2015.
https://www.bts.gov/content/miles-freight-railroad-operated-class-railroad

Cadena, C., L. Carlone, H. Carillo, Y. Latif, D. Scaramuzza, J. Neira, I. Reid, and J. J. Leonard, "Past, Present, and Future of Simultaneous Localization and Mapping: Toward the Robust-Perception Age," *IEEE Transactions in Robotics*, Vol. 32, 2016, pp. 1309–1332.

Cameron, Alan, "The Inner Edge: Who Holds the Key to Indoor Nav?" *GPS World*, April 16, 2013. As of October 13, 2019:
http://gpsworld.com/the-inner-edge-who-holds-the-key-to-indoor-nav/

Carrington, R. C., "Description of a Singular Appearance Seen in the Sun on September 1, 1859," *Monthly Notices of the Royal Astronomical Society*, Vol. 20, 1859, pp. 13–15.

Carroll, James, and Kirk Montgomery, "Global Positioning System Timing Criticality Assessment—Preliminary Performance Results," *Proceedings of the 40th Annual Precise Time and Time Interval Systems and Applications Meeting*, Reston, Va.: ION Publications, 2008. As of October 13, 2019:
https://rntfnd.org/wp-content/uploads/2013/09/GPS-Timing-Criticality-Volpe-Paper-2008.pdf

Carroll, Kevin, Anthony Hawes, Benjamin Peterson, Kenneth Dykstra, Peter Swaszek, and Sherman Lo, "Differential Loran-C," *Proceedings of the European Navigation Conference—GNSS 2004*, Rotterdam, The Netherlands, May 2004.

Cartlidge, Edwin, "Atom Interferometry Heats Up with Warm-Vapour Device," *Physics World*, May 1, 2017. As of October 13, 2019:
https://physicsworld.com/a/atom-interferometry-heats-up-with-warm-vapour-device/

Cavitt, S. A., D. R. Halla, T. D. Holman, S. L. Patterson, and C. R. Swain, *Precision Timing Requirements Summary: Communications Sector, Electricity Subsector, Emergency Services Sector, and Financial Services Sector*, Laurel, Md.: Johns Hopkins University Applied Physics Laboratory, AOS-17-0771, January 2018.

Celano, T. P., B. B. Peterson, and C. A. Schue, "Low Cost Digitally Enhanced Loran for Tactical Applications (LC DELTA)," *Proceedings of the 33rd Annual Convention and Technical Symposium*, Tokyo: International Loran Association, October 2004.

Chairman of the Council of Economic Advisers, *Economic Report of the President*, Washington, D.C., February 2018. As of October 13, 2019:
https://www.govinfo.gov/app/collection/ERP

Chipka, J. B., and M. Campbell, "Autonomous Urban Localization and Navigation with Limited Information," *Proceedings of the Intelligent Vehicles Symposium (IV), 2018 IEEE*, Piscataway, N.J.: IEEE, arXiv:1810.04243 [cs.RO], 2018.

Chronos Technology, "JammerCam," webpage, 2018. As of October 13, 2019:
http://www.gps-world.biz/products/gnss-interference-detection/jammercam

Clark, Greg, "UK Space Agency Leads Work on Options for Independent Satellite System," GOV.UK, May 2, 2018. As of October 13, 2019:
https://www.gov.uk/government/news/
uk-space-agency-leads-work-on-options-for-independent-satellite-system

"Coast Guard to Discontinue Service from Remaining Differential GPS Sites," *Coast Guard News*, March 23, 2018. As of October 13, 2019:
http://coastguardnews.com/
coast-guard-to-discontinue-service-from-remaining-differential-gps-sites/2018/03/23/

Cobb, James C., "The Value of Geologic Maps and the Need for Digitally Vectorized Data," in *Workshop Proceedings of Digital Mapping Techniques '02*, ed. D. R. Soller, Salt Lake City, Utah: U.S. Geological Survey Open-File Report 02-370, 2002. As of October 13, 2019:
https://pubs.usgs.gov/of/2002/of02-370/cobb.html

Coffed, Jeff, *The Threat of GPS Jamming: The Risk to an Information Utility*, Melbourne, Fla.: Harris Corporation, 2016.

Communications Security, Reliability, and Interoperability Council and Working Group 3, *E9-1-1 Location Accuracy: Indoor Location Test Bed Report*, Washington, D.C.: Federal Communications Commission, March 14, 2013. As of October 13, 2019:
https://transition.fcc.gov/bureaus/pshs/advisory/csric3/
CSRIC_III_WG3_Report_March_%202013_ILTestBedReport.pdf

Complementary Positioning, Navigation, and Timing Capability; Notice; Request for Public Comments, 80 Fed. Reg. 15268, March 23, 2015.

Compton, R., K. Nelson, and C. Fertig, "Systems and methods for eliminating multi-path errors from atomic inertial sensors," US Patent US9887019B2, filed February 4, 2016, and issued February 6, 2018.

Cosart, Lee, and Marc Weiss, "Wide-Area Time Distribution with PTP Using Commercial Telecom Optical Fiber," *NASPI Work Group Meeting*, Gaithersburg, Md.: North American SynchroPhasor Initiative, March 22, 2017.

Costin, Andrei, and Aurélien Francillon, "Ghost in the Air(Traffic): On Insecurity of ADS-B Protocol and Practical Attacks on ADS-B Devices," *Black Hat USA 2012*, Las Vegas, Nev., 2012.

Craig, Desiree, Derek Ruff, Steve Hewitson, Joel Barnes, and John Amt, "The UHARS Non-GPS Based Positioning System," *Proceedings of the 25th International Technical Meeting of the Satellite Division of the Institute of Navigation (ION GNSS 2012)*, Nashville, Tenn.: ION Publications, September 2012. As of October 13, 2019:
http://www.locata.com/wp-content/uploads/2012/09/ION-2012_NGBPS-Paper-Final-from -USAF.pdf

CSRIC—*See* Communications Security, Reliability, and Interoperability Council.

Curry, Charles, "Dependency of Communications Systems on PNT Technology," Gloucestershire: Chronos Technology Ltd., May 25, 2010. As of October 13, 2019:
https://www.chronos.co.uk/files/pdfs/wps/Dependency_of_Comms_on_PNT_Technology.pdf

———, "GPS Jamming—Quantifying the Threat," presentation at Workshop on Synchronization and Timing Systems, San Jose, Calif., April 16–18, 2013. As of October 13, 2019:
https://tf.nist.gov/seminars/WSTS/PDFs/7-2_CTL_Curry_GPS_Jamming%20ver%202.pdf

———, "Detecting GPS Jammers 'Gone in 20 Seconds,'" presentation at 54th Meeting of the Civil GPS Service Interface Committee, Tampa, Fla., September 8, 2014. As of October 13, 2019:
https://www.gps.gov/cgsic/meetings/2014/curry.pdf

Cushman & Wakefield, "Top U.S. Container Ports—Ocean Freight Volume" (table), 2019. As of October 13, 2019:
http://blog.cushwake.com/wp-content/uploads/2019/03/Top-US-Seaports-YE-2018.pdf

Dahlen, A. P., "A Prototype Cesium Clock Ensemble for the Loran-C Radionavigation System," *40th Annual Precise Time and Time Interval (PTTI) Meeting*, Manassas, Va.: ION Publications, 2008, pp. 227–240.

DARPA—*See* Defense Advanced Research Projects Agency.

DeCorla-Souza, Patrick, Douglass Lee, Darren Timothy, and Jennifer Mayer, "Comparing Public-Private Partnerships with Conventional Procurement: Incorporating Considerations from Benefit-Cost Analysis," *Transportation Research Journal*, No. 2346, 2013, pp. 32–39.

Defense Advanced Research Projects Agency, "Beyond GPS: 5 Next-Generation Technologies for Positioning, Navigation & Timing (PNT)," press release, Alexandria, Va., 2014. As of October 13, 2019:
https://www.darpa.mil/news-events/2014-07-24

DHS—*See* U.S. Department of Homeland Security.

Dignan, Larry, "5G Takes Center Stage at CES 2018 with Actual Deployments Later," ZDNet, January 8, 2018. As of October 13, 2019:
https://www.zdnet.com/article/5g-takes-center-stage-at-ces-2018-with-actual-deployments-later/

DoD—*See* U.S. Department of Defense.

Doherty, James, *Independent Assessment Team (IAT) Report on Findings and Recommendations on WAAS Network Time (WNT)*, Alexandria, Va.: Institute for Defense Analysis, 2011.

DoT—*See* U.S. Department of Transportation.

Dotinga, Randy, "Fact Check: Mysterious Outage Unleashes S.D. Chaos?" Voice of San Diego, June 10, 2011. As of October 13, 2019:
https://www.voiceofsandiego.org/topics/news/fact-check-mysterious-outage-unleashes-s-d-chaos/

Durrant-Whyte, H., and T. Bailey, "Simultaneous Localization and Mapping: Part 1," *IEEE Robotics & Automation Magazine*, June 2006.

Eidson, John, "IEEE-1588 Standard for a Precision Clock Synchronization Protocol for Networked Measurement and Control Systems: A Tutorial," presentation at National Institute of Standards and Technology, Gaithersburg, Md., October 10, 2005. As of October 13, 2019:
https://www.nist.gov/sites/default/files/documents/el/isd/ieee/tutorial-basic.pdf

Emerson, Sarah, "FEMA Is Preparing for a Solar Superstorm That Would Take Down the Grid," Motherboard, June 20, 2017. As of October 13, 2019:
https://motherboard.vice.com/en_us/article/ev473k/fema-is-preparing-for-a-solar-superstorm-that-would-take-down-the-grid

Empson, Rip, "Apple Acquires Indoor GPS Startup WiFiSlam for $20 Million," TechCrunch, March 24, 2013. As of October 13, 2019:
https://techcrunch.com/2013/03/24/apple-acquires-indoor-gps-startup-wifislam-for-20m/

Engler, E., M. Hoppe, J. Ritterbusch, T. Ehlers, C. Becker, K.-C. Ehrke, and H. Callsen-Bracker, "Guidelines for the Coordinated Enhancement of the Maritime Position, Navigation and Time Data System," *Scientific Journals of the Maritime University of Szczecin*, Vol. 45, No. 117, 2016, pp. 44–53.

Engler, E., T. Noack, M. Hoppe, R. Ziebold, and Z. Dai, *Resilient PNT: Vision and Mission*, Oslo: e-Navigation Underway, 2012.

English, Gordon, David Hackston, John Greenway, and Randolph Helland, *Safety Profile of the Great Lakes-St. Lawrence Seaway System*, Glenburnie, Ont.: Research and Traffic Group, March 2014. As of October 13, 2019:
https://www.seaway.dot.gov/sites/seaway.dot.gov/files/docs/Safety%20Profile%20-%20Full%20Report.pdf

Erkut, Erhan, Stevanus A. Tjandra, and Vedat Verter, "Hazardous Materials Transportation," in Cynthia Barnhart and Gilbert Laporte, eds., *Handbooks in Operations Research and Management Science*, Oxford, U.K.: Elsevier, 2007, p. 541.

European Global Navigation Satellite Systems Agency, *GNSS User Technology Report*, Luxembourg: Publications Office of the European Union, No. 1, 2016. As of October 13, 2019:
https://www.gsa.europa.eu/european-gnss/gnss-market/2016-gnss-user-technology-report

———, *GNSS Market Report*, No. 5, 2017.

———, "World's First Dual-Frequency GNSS Smartphone Hits the Market," press release, Saint-Germain-en-Laye, France, June 4, 2018. As of October 13, 2019:
https://www.gsa.europa.eu/newsroom/news/world-s-first-dual-frequency-gnss-smartphone-hits-market

European GNSS Agency—*See* European Global Navigation Satellite Systems Agency.

European Union, "Commission Delegated Regulation (EU) 2017/574, supplementing Directive 2014/65/EU of the European Parliament and of the Council with regard to regulatory technical standards for the level of accuracy of business clocks," June 7, 2016. As of August 10, 2020:
https://eur-lex.europa.eu/legal-content/EN/TXT/PDF/?uri=CELEX:32017R0574&from=EN

Eyer, Jim, and Garth Corey, *Energy Storage for the Electricity Grid: Benefits and Market Potential Assessment Guide*, Albuquerque, N.M.: Sandia National Laboratories, 2010.

FAA—*See* Federal Aviation Administration.

FAO—*See* Food and Agriculture Organization of the United Nations.

FCC—*See* Federal Communications Commission.

Federal Aviation Administration, *The Future of the NAS*, Washington, D.C.: NextGen, June 2016a. As of October 13, 2019:
https://www.faa.gov/nextgen/media/futureofthenas.pdf

———, "ADS-B: Wide Area Multilateration (WAM)," webpage, October 27, 2016b. As of December 11, 2018:
https://www.faa.gov/nextgen/programs/adsb/atc/wam/

Federal Communications Commission, "Jammer Enforcement," webpage, 2018. As of October 13, 2019:
https://www.fcc.gov/general/jammer-enforcement

Federal Highway Administration, "Popular Vehicle Trips Statistics: Number of Vehicle Trips (VT) by Household Income," *2017 National Household Travel Survey*, Washington, D.C.: U.S. Department of Transportation, 2017. As of October 13, 2019:
https://nhts.ornl.gov/vehicle-trips

Federal Reserve Bank of St. Louis, "Total Gross Domestic Product for Los Angeles-Long Beach-Anaheim, CA MSA) (NGMP31080)," undated. As of October 13, 2019:
https://fred.stlouisfed.org/series/NGMP31080

Fernholz, Tim, "The Entire Global Financial System Depends on GPS, and It's Shockingly Vulnerable to Attack," Quartz, October 22, 2017. As of October 13, 2019:
https://qz.com/1106064/the-entire-global-financial-system-depends-on-gps-and-its-shockingly -vulnerable-to-attack/

Financial Industry Regulatory Authority Compliance Audit Trail, "Standards for Self Reporting Deviations of Clock Synchronization Standards to FINRA CAT," CAT Alert 2020-02, Version 1.1, May 8, 2020. As of August 10, 2020:
https://www.catnmsplan.com/sites/default/files/2020-05/CAT-Alert-2020-02-v1.1.pdf

FirstNet: First Responder Network Authority, "Top 10 Frequently Asked Questions," undated. As of October 13, 2019:
https://firstnet.gov/sites/default/files/Top_Ten_FAQs_180514.pdf

Fleishman, Glenn, "How the iPhone Knows Where You Are," Macworld, April 28, 2011. As of October 13, 2019:
https://www.macworld.com/article/1159528/smartphones/how-iphone-location-works.html

Food and Agriculture Organization of the United Nations, "Fishery and Aquaculture Country Profiles: The Republic of Korea," webpage, June 2017. As of October 13, 2019:
http://www.fao.org/fishery/facp/KOR/en

Frame, Jim, "Do You Pay Attention to the Space Weather Warnings: Response," RPLS Today, September 12, 2014. As of October 13, 2019:
https://rplstoday.com/community/gnss-geodesy/do-you-pay-attention-to-the-space-weather-warnings/

Gallagher, S., "Radio Navigation Set to Make Global Return as GPS Backup, Because Cyber," Ars Technica, August 7, 2017. As of October 13, 2019:
https://arstechnica.com/gadgets/2017/08/radio-navigation-set-to-make-global-return-as-gps -backup-because-cyber/

GAO—*See* U.S. Government Accountability Office.

Gates, C., and G. Pattabiraman, "Terrestrial Beacons Bring Wide-Area Location Indoors," GPS World, July 12, 2016. As of October 13, 2019:
https://www.researchgate.net/publication/320777202_Terrestrial_beacons_bring_wide-area_location_indoors

Gauthier, J. P., E. P. Glennon, C. C. Rizos, and A. G. Dempster, "Time Transfer Performance of Locata—Initial Results," presentation at the Institute of Navigation Precise Time and Time Interval (PTTI) Conference, Seattle, Wash., December 2–5, 2013.

Gedik, Ridvan, Hugh Medal, Chase Rainwater, Ed A. Pohl, and Scott J. Mason, "Vulnerability Assessment and Re-routing of Freight Trains under Disruptions: A Coal Supply Chain Network Application," *Transportation Research Part E: Logistics and Transportation Review*, Vol. 71, 2014.

Gill, Alexander, Steven J. Byrnes, Jennifer Choy, Christine Y. Wang, Matthew A. Sinclair, Adam Kelsey, and David Johnson, "Cold Atom Interferometry," US Patent US20170372808A1, filed June 22, 2016, and issued December 18, 2018.

Gill, Paul, "Tactical Innovation and the Provisional Irish Republican Army," *Studies in Conflict & Terrorism*, Vol. 40, No. 7, 2017, pp. 573–585.

Gill, Paul, John Horgan, Samuel T. Hunter, and Lily D. Cushenbery, "Malevolent Creativity in Terrorist Organizations," *Journal of Creative Behavior*, Vol. 47, No. 2, 2013, pp. 125–151.

Glass, Dan, "What Happens If GPS Fails?" *The Atlantic*, June 13, 2016. As of October 13, 2019:
https://www.theatlantic.com/technology/archive/2016/06/what-happens-if-gps-fails/486824/

Global Mobile Suppliers Association, *Evolution to LTE Report*, Farnham, U.K.: GSA, October 26, 2016.

Gonzaga, Shireen, ed., *Hubble Space Telescope Primer for Cycle 21*, Baltimore, Md.: Space Telescope Science Institute, December 2012. As of October 13, 2019:
http://www.stsci.edu/itt/review/cp_primer_cy21/cp_primer/primer.pdf

Goode, Thomas, "Letter to J. Platt, RE: Meeting Summary of ATIS/DHS Call on December 8, 2017," Washington, D.C.: Alliance for Telecommunications Industry Solutions, January 26, 2018. As of October 13, 2019:
http://atis.org/01_legal/docs/SYNC/ATIS_SYNC_Letter_to_DHS_1-26-18.pdf

Google, Google Earth Pro, version 7.3, 2019.

GPS World Staff, "NextNav and Broadcom Partner for Indoor Accuracy," GPS World, October 21, 2013.

———, "NextNav Supports Metropolitan Beacon System for Mobile," GPS World, January 19, 2016.

———, "Iridium Constellation Provides Low-Earth Orbit Satnav Service," GPS World, January 12, 2017.

GPS.gov, "Space Segment," webpage, March 21, 2019. As of October 13, 2019:
https://www.gps.gov/systems/gps/space/

Graham, W., "Iridium NEXT-5 Satellites Ride to Orbit on SpaceX Falcon 9," NASA Spaceflight.com, March 29, 2018. As of October 13, 2019:
https://www.nasaspaceflight.com/2018/03/iridium-next-5-satellites-spacex-falcon-9/

Grant, Alan, Paul Williams, Nick Ward, and Sally Basker, "GPS Jamming and the Impact on Maritime Navigation," *Journal of Navigation*, Vol. 62, No. 2, 2009.

Gray, J., and D. P. Siewiorek, "High-availability Computer Systems," *Computer*, Vol. 24, No. 9, 1991, pp. 39–48.

Grinberg, Emanuella, Jon Ostrower, Madison Park, and Christina Zdanowicz, "Atlanta's Hartsfield-Jackson Airport Restores Power after Crippling Outage," CNN, December 18, 2017. As of October 13, 2019:
https://www.cnn.com/2017/12/17/us/atlanta-airport-power-outage/index.html

Grisso, Robert Dwight, Marcus M. Alley, and Gordon Eugene Groover, "Precision Farming Tools: GPS Navigation," Virginia Cooperative Extension, Publication 442-501, May 26, 2005. As of October 13, 2019:
https://vtechworks.lib.vt.edu/bitstream/handle/10919/51374/442-501.pdf

Groover, Gordon, and Robert Grisso, "Investing in GPS Guidance Systems?" Virginia Cooperative Extension, Publication 448-076, 2009. As of June 26, 2018:
https://pubs.ext.vt.edu/448/448-076/448-076.html

GSA—*See* Global Mobile Suppliers Association.

Gutt, Gregory, David Lawrence, Stewart Cobb, and Michael O'Connor, "Recent PNT Improvements and Test Results Based on Low Earth Orbit Satellites," *Proceedings of the 2018 International Technical Meeting of the Institute of Navigation*, Reston, Va.: ION Publications, January 2018, pp. 570–577.

Gutt, Gregory, and Michael O'Connor, "Trusted Time and Location, Everywhere," presentation at the 17th National Space-Based Positioning, Navigation, and Timing (PNT) Advisory Board Meeting, May 18–19, 2016. As of October 13, 2019:
https://www.gps.gov/governance/advisory/meetings/2016-05/gutt.pdf

Hall, Peter V., "'We'd Have to Sink the Ships': Impact Studies and the 2002 West Coast Port Lockout," *Economic Development Quarterly*, Vol. 18, No. 4, 2004.

Halsing, David, Kevin Theissen, and Richard Bernkopf, "A Cost-Benefit Analysis of *The National Map*," U.S. Geological Survey Circular 1271, November 5, 2004.

Hambling, David, "GPS Chaos: How a $30 Box Can Jam Your Life," *New Scientist*, March 4, 2011. As of October 13, 2019:
https://www.newscientist.com/article/dn20202-gps-chaos-how-a-30-box-can-jam-your-life/

———, "Ships Fooled in GPS Spoofing Attack Suggest Russian Cyberweapon," *New Scientist*, August 10, 2017.

Hanemann, W. Michael, "Willingness to Pay and Willingness to Accept: How Much Can They Differ?" *The American Economic Review*, Vol. 81, No. 3, 1991, pp. 635–647.

Hastings, D., and H. Garrett, *Spacecraft-Environment Interactions*, Cambridge, U.K.: Cambridge University Press, 1996.

Helwig, A., G. Offermans, C. Stout, and C. Schue, "eLoran System Definition and Signal Specification Tutorial," presentation at UrsaNav briefing at International Loran Association (ILA-40), Busan, South Korea, November 2011.

Hodgson, R., "On a Curious Appearance Seen in the Sun," *Monthly Notices of the Royal Astronomical Society*, Vol. 20, 1859, pp. 15–16.

Horowitz, Michael C., Evan Perkoski, and Philip B. K. Potter, "Tactical Diversity in Militant Violence," *International Organization*, Vol. 72, Winter 2018, pp. 139–171.

Hsu, Jeremy, "U.S. 'Master Clock' Keepers Test Terrestrial Alternative to GPS," *IEEE Spectrum*, September 22, 2015. As of October 13, 2019:
https://spectrum.ieee.org/tech-talk/telecom/wireless/us-master-clock-keepers-test-ground-alternative-to-gps

Huang, Ronald K., and Apple Inc., "Determining a location of a mobile device using a location database," U.S. Patent 8.700060B2, filed on January 15, 2010, and issued April 15, 2014.

Humphreys, Todd, "Statement on the Vulnerability of Civil Unmanned Aerial Vehicles and Other Systems to Civil GPS Spoofing," testimony presented to the Subcommittee on Oversight, Investigations, and Management of the House Committee on Homeland Security, July 18, 2012.

Incheon Port Authority, "Incheon Port Handles 223,000 TEU in April," press release, Incheon, South Korea, May 30, 2016. As of December 31, 2019:
https://www.icpa.or.kr/eng/article/view.do?articleKey=9141&searchSelect=title&searchValue=223%2C000+TEU&boardKey=266&menuKey=623&currentPageNo=1

Institute for Economics and Peace, "Financing Terror," in *Global Terrorism Index 2017: Measuring and Understanding the Impact of Terrorism*, New York: IEP, 2017. As of June 26, 2018:
http://visionofhumanity.org/app/uploads/2017/11/Global-Terrorism-Index-2017.pdf

International Loran Association, *Enhanced Loran (eLoran) Definition Document*, Santa Barbara, Calif.: ILA, January 12, 2007.

Irani, Darius, Michael Siers, Catherine Menking, and Zachary Nickey, *Economic and Fiscal Impact Analysis of Class I Railroads in 2017*, Towson, Md.: Regional Economic Studies Institute, 2018.

Iridium, "Iridium Launches Breakthrough Alternative Global Positioning System (GPS) Service," press release, McLean, Va., May 23, 2016. As of October 13, 2019:
https://www.globenewswire.com/news-release/2016/05/23/842381/0/en/Iridium-Launches-Breakthrough-Alternative-Global-Positioning-System-GPS-Service.html

Jackson, Brian A., John C. Baker, Peter Chalk, Kim Cragin, John V. Parachini, and Horacio R. Trujillo, *Aptitude for Destruction*, Volume 1: *Organizational Learning in Terrorist Groups and Its Implications for Combating Terrorism*, Santa Monica, Calif.: RAND Corporation, MG-331-NIJ, 2005a. As of August 26, 2019:
https://www.rand.org/pubs/monographs/MG331.html

———, *Aptitude for Destruction*, Volume 2: *Case Studies of Organizational Learning in Five Terrorist Groups*, Santa Monica, Calif.: RAND Corporation, MG-332-NIJ, 2005b. As of August 26, 2019:
https://www.rand.org/pubs/monographs/MG332.html

Jackson, Brian A., and David R. Frelinger, "Rifling Through the Terrorists' Arsenal: Exploring Groups' Weapon Choices and Technology Strategies," *Studies in Conflict & Terrorism*, Vol. 31, No. 7, 2008, pp. 583–604.

———, *Understanding Why Terrorist Operations Succeed or Fail*, Santa Monica, Calif.: RAND Corporation, OP-257-RC, 2009. As of August 26, 2019:
https://www.rand.org/pubs/occasional_papers/OP257.html

Jaldell, Henrik, *Tidsfaktorns betydelse vid räddningsinsatser—en uppdatering av en samhällseknomisk studie*, Karldysf: Swedish Rescue Service Agency, Report P21-499/04, 2004. As of October 13, 2019:
https://www.msb.se/RibData/Filer/pdf/19958.pdf

Jaldell, Henrik, Prachaksvich Lebnak, and Anurak Amornpetchsathaporn, "Time Is Money, but How Much? The Monetary Value of Response Time for Thai Ambulance Emergency Services," *Value in Health*, Vol. 17, No. 5, 2014, pp. 555–560.

Jane's by IHS Markit, "SCL-300 GPS Jammer," March 13, 2016.

———, "Optima-2.2 GPS and GLONASS System Navigation Jammer," August 15, 2017a.

———, "Optima-B GPS Terminal Spoofing Jammer," August 16, 2017b.

———, "GPSJ-25 GPS Jammer," September 15, 2017c.

———, "GPSJ-40 GPS Jammer," September 15, 2017d.

———, "GPSJ-50 GPS Jammer," September 15, 2017e.

Jansen, Bart, "Delta: Atlanta Airport Power Outage Cost $25M to $50M in Income," *USA Today*, January 3, 2018a.

———, "Delta Lost $60 Million in December from Atlanta Airport Power Outage, Winter Storm," *USA Today*, January 11, 2018b.

Jewell, Don, "Iridium and GPS Revisited: A New PNT Solution on the Horizon?" *GPS World*, June 8, 2016.

Jiang, Yi, Shufang Zhang, and Junhua Teng, "An Analysis of Positioning Method in AIS Ranging Mode," *Automatic Control and Computer Sciences*, Vol. 52, No. 4, July 2018, pp. 322–333.

Jiggens, Piers, Marc-Andre Chavy-Macdonald, Giovanni Santin, Alessandra Menicucci, Hugh Evans, and Alain Hilgers, "The Magnitude and Effects of Extreme Solar Particle Events," *Journal of Space Weather and Space Climate*, Vol. 4, 2014, A20. As of October 13, 2019: https://www.swsc-journal.org/articles/swsc/pdf/2014/01/swsc130038.pdf

John A. Volpe National Transportation Systems Center, *Vulnerability Assessment of the Transportation Infrastructure Relying on the Global Positioning System*, Cambridge, Mass.: U.S. Department of Transportation, 2001.

———, *Benefit-Cost Assessment Refresh: The Use of* eLORAN *to Mitigate GPS Vulnerability for Positioning, Navigation, and Timing Services: Final Report*, Cambridge, Mass.: U.S. Department of Transportation, November 5, 2009. As of October 13, 2019: https://rntfnd.org/wp-content/uploads/Benefit-Cost-of-eLoran-Volpe-Center-2009.pdf

John Deere, "John Deere Guidance Systems: Guidance You Can Grow With," brochure, Moline, Ill., 2018. As of October 13, 2019: https://www.deere.com/common/docs/products/equipment/agricultural_management_solutions/guidance_systems/brochure/en_GB_yy1114823_e.pdf

Johnson, David, David Butts, Richard Stoner, and Tom Thornvalden, "Systems and methods for multiple species atom interferometry," U.S. Patent WO2016069341A1, filed October 31, 2014, and issued May 6, 2016.

Johnson, Gregory, and Peter Swaszek, *Feasibility Study of R-Mode Combining MF DGNSS, AIS, and eLoran Transmissions*, Brussels: ACCSEAS Project, European Regional Development Fund, September 25, 2014. As of October 13, 2019: http://www.accseas.eu/content/downloadstream/7206/65272/Feasibility%2520Study%2520of%2520R-Mode%2520combining%2520MF%2520DGNSS%2520AIS%2520and%2520eLoran%2520Transmissions.pdf

Johnson, G. W., P. F. Swaszek, R. J. Hartnett, R. Shalaev, and M. Wiggins, "An Evaluation of eLoran as a Backup to GPS," paper presented at the 2007 IEEE Conference on Technologies for Homeland Security, Woburn, Mass., May 16–17, 2007.

Johnson, R. C., "Indoor Nav Leverages LTE: Nanosecond Beacons Boost E911," *EE Times*, August 30, 2016. As of October 13, 2019: https://www.eetimes.com/document.asp?doc_id=1330379

Jones, Michael, "GNSS Protection Overview 2017," presentation at the 20th Meeting of the Space-Based Positioning, Navigation, and Timing National Advisory Board, Redondo Beach, Calif., November 15, 2017. As of October 13, 2019: https://www.gps.gov/governance/advisory/meetings/2017-11/jones.pdf

Josephs, Leslie, "Delta: Atlanta Power Outage Cost It up to $50 Million," *CNBC*, January 3, 2018.

Kacem, Thabet, Alexandre Barreto, Duminda Wijesekera, and Paulo Costa, "ADS-Bsec: A Novel Framework to Secure ADS-B," *ICT Express*, Vol. 3, No. 4, 2017, pp. 160–163.

Kaur, Namneet, Florian Frank, Paul-Eric Pottie, and Philip Tuckey, "Time Transfer over a White Rabbit Network," briefing at Facilities for Innovation, Research, Services and Training in Time and Frequency (FIRST-TF) General Assembly, l'Institut d'Optique d' Aquitaine, Talence, France, June 8, 2017a. As of October 13, 2019: http://first-tf.fr/wp-content/uploads/2017/06/FIRSTTF_AG2017_Doctorat_NamneetKaur.pdf

———, "Time and Frequency Transfer over a 500 km Cascaded White Rabbit Network," presentation at the 2017 Joint Conference of the European Frequency and Time Forum and IEEE International Frequency Control Symposium (EFTF/IFCS), Besancon, France, July 9–13, 2017b.

Kawecki, J., J. Brewer, J. Cao, and J. Baldwin, "Initial Operational Capability for the Ultra High-Accuracy Reference System," *GPS World*, August 2016, pp. 22–29.

Keller, John, "NASA Approaches Industry for Radiation-hardened ASIC Designs Able to Withstand Rigors of Mission to Jupiter," *Military & Aerospace Electronics*, January 1, 2011. As of October 13, 2019: https://www.militaryaerospace.com/computers/article/16723608/nasa-approaches-industry-for -radiationhardened-asic-designs-able-to-withstand-rigors-of-mission-to-jupiter

Kelley, Chris, Joan Pellegrino, and Emmanuel Taylor, *Time Distribution Alternatives for the Smart Grid, Workshop Report*, Gaithersburg, Md.: National Institute of Standards and Technology, NIST Special Publication 1500-12, November 2017.

Kennedy, Robert A., "GPS Time Synchronization," *EC&M*, August 19, 2011. As of October 13, 2019: http://www.ecmweb.com/computers-amp-software/gps-time-synchronization

Kettle, Louise, and Andrew Mumford, "Terrorist Learning: A New Analytical Framework," *Studies in Conflict & Terrorism*, Vol. 40, No. 7, 2017, pp. 523–538.

Knight, Sarah E., Carys Keane, and Amy Murphy, "Adversary Group Decision-Making Regarding Choice of Attack Methods: Expecting the Unexpected," *Terrorism and Political Violence*, Vol. 29, No. 4, 2017, pp. 713–734.

Kim, C., H. So, T. Lee, and C. Kee, "A Pseudolite-Based System for Legacy GNSS Receivers," *Sensors*, Vol. 14, 2014, pp. 6104–6123.

Kim, Hyunh-Jin, "Seoul: North Korea Fires Missile, Tries to Jam GPS Signals," *Military Times*, April 1, 2016. As of October 13, 2019: https://www.militarytimes.com/news/your-military/2016/04/01/ seoul-north-korea-fires-missile-tries-to-jam-gps-signals/

Kim, Jack, and Jonathan Saul, "South Korea Revives GPS Backup Project After Blaming North for Jamming," Reuters, May 1, 2016. As of October 13, 2019: https://www.reuters.com/article/us-shipping-southkorea-navigation/south-korea-revives-gps -backup-project-after-blaming-north-for-jamming-idUSKCN0XT01T

Kitching, J., "NIST Chip-Scale Atomic Device Program," NIST briefing, undated.

Klecka, Rudy, "Fundamentals of Precision Time Protocol," presentation at the ODVA 2015 Industry Conference, Frisco, Tex., October 13–15, 2015.

Koivisto, M., Al Hakkarainen, M. Costa, P. Kela, K. Leppanen, and M. Valkama, "High Efficiency Device Positioning and Location-Aware Communications in Dense 5G Networks," *IEEE Communications Magazine*, Vol. 55, No. 8, July 2017.

Krausmann, Elisabeth, Emmelie Andersson, Mark Gibbs, and William Murtagh, *Space Weather & Critical Infrastructures: Findings and Outlook*, Luxembourg: Publications Office of the European Union, JRC104231, 2016.

Kweun, Jeong Yun, Porter K. Wheeler, and Jonathan L. Gifford, "Evaluating Highway Public-Private Partnerships: Evidence from U.S. Value for Money Studies," *Transport Policy*, Vol. 62, 2018, pp. 12–20.

Lammers, Marre, Wim Bernascom, and Henk Elffers, "How Long Do Offenders Escape Arrest? Using DNA Traces to Analyse when Serial Offenders Are Caught," *Journal of Investigative Psychology and Offender Profiling*, Vol. 9, 2012, pp. 13–29.

Langer, J., J. Clark, and T. Powell, "Transit: The GPS Forefather," *Crosslink Magazine* (The Aerospace Corporation), April 1, 2010.

Lawrence, David, H. S. Cobb, G. Gutt, M. O'Connor, T. G. R. Reid, T. Walter, and D. Whelan, "Innovation: Navigation from LEO," *GPS World*, June 30, 2017.

Leveson, Irv, *Socio-Economic Benefits Study: Scoping the Value of CORS and GRAV-D*, Jackson, N.J.: Leveson Consulting, NCNL0000-8-37007, 2009.

———, "Recognizing GPS Contributions," presentation to the National Space-Based Positioning, Navigation, and Timing Advisory Board, Annapolis, Md., May 7, 2013. As of October 13, 2019: https://www.gps.gov/governance/advisory/meetings/2013-05/leveson.pdf

———, "The Economic Value of GPS: Preliminary Assessment," presentation to the National Space-Based Positioning, Navigation, and Timing Advisory Board, Annapolis, Md., June 11, 2015a. As of October 13, 2019: https://www.gps.gov/governance/advisory/meetings/2015-06/leveson.pdf

———, Irv, *GPS Civilian Economic Value to the U.S., Interim Report*, Greenbelt, Md.: ASRC Federal Research and Technology Solutions, August 31, 2015b. As of October 13, 2019: http://www.performance.noaa.gov/wp-content/uploads/2015-08-31-Phase-1-Report-on-GPS-Economic-Value.pdf

Li, Han, Liuyan Han, Ran Duan, and Geoffrey M. Garner, "Analysis of the Synchronization Requirements of 5G and Corresponding Solutions," *IEEE Communications Standards Magazine*, March 2017.

Locata, "FAQs," webpage, undated. As of October 13, 2019: http://www.locata.com/technology/faqs/

———, "Technology Brief," Canberra: Locata Technology Brief v8.0, July 2014. As of October 13, 2019: http://www.locata.com/wp-content/uploads/2014/07/Locata-Technology-Brief-v8-July-2014-Final1.pdf

Lombardi, Michael A., *NIST Time and Frequency Services*, 2002 ed., Gaithersburg, Md.: National Institute of Standards and Technology, NIST Special Publication 432, April 2003.

Lombardi, Michael A., Andrew N. Novick, George Neville-Neil, and Ben Cooke, "Accurate, Traceable, and Verifiable Time Synchronization for World Financial Markets," *Journal of Research of the National Institute of Standards and Technology*, Vol. 121, 2016, pp. 436–463.

Long Range Aids to Navigation (LORAN) Program; Office of Navigation and Spectrum Management, 72 Fed. Reg. 796, January 8, 2007.

Lowe, J., M. Deutch, G. Nelson, D. Sutton, W. Yates, P. Hansen, O. Eliezer, T. Jung, S. Morrison, Y. Liang, and D. Rajan, "New Improved System for WWVB Broadcast," in *Proceedings of the 43rd Annual Precise Time and Time Interval Systems and Applications Meeting*, Reston, Va.: ION Publications, November 2011, pp. 163–184.

Lumbreras, Cristina, and Gary Machado, "112 Caller Location & GNSS," presentation at the European Emergency Number Association, Ref. Ares(2014)1665619, May 22, 2014.

Maddison, Ralph, and Cliona Ni Mhurchu, "Global Positioning System: A New Opportunity in Physical Activity Measurement," *International Journal of Behavioral Nutrition and Physical Activity*, Vol. 6, No. 73, 2009.

Marais, Juliette, Julie Beugin, and Marion Berbineau, "A Survey of GNSS-Based Research and Developments for the European Railway Signaling," *IEEE Transactions on Intelligent Transportation Systems*, Vol. 18, No. 10, 2017.

Martin, Steven W., James Hanks, Aubrey Harris, Gene Wills, and Swagata Banerjee, "Estimating Total Costs and Possible Returns from Precision Farming Practices," *Crop Management*, Vol. 4, No. 1, 2005. As of October 13, 2019:
https://pubag.nal.usda.gov/pubag/downloadPDF.xhtml?id=11889&content=PDF

Martinez, Carlos E., Gerry McNeill, Jill Kamienski, Nicholas Kasdaglis, Y. Ryan Kuo, Yueh Quach, Brian Bian, Jianming She, and Young C. Lee, *Alternate Position, Navigation, and Timing Knowledge Elicitation Report*, McLean, Va.: The MITRE Corporation, 2017.

Masunaga, S., "GPS Guidance Can Be Fooled, So Researchers Are Scrambling to Find Backup Technologies," *Los Angeles Times*, March 15, 2018. As of October 13, 2019:
http://www.latimes.com/business/la-fi-gps-alternatives-20180315-story.html

Mazzocchi, Mario, Francesca Hansstein, and Maddalena Ragona, "The 2010 Volcanic Ash Cloud and Its Financial Impact on the European Airline Industry," *CESifo Forum*, Vol. 11, No. 2, June 2010, pp. 92–100.

McCallie, Donald L., *Exploring Potential Ads-B Vulnerabilites in the FAA's Nextgen Air Transportation System*, Wright-Patterson Air Force Base, Ohio: Air University, 2011.

McNichol, Dan, *The United States: The World's Largest Emerging P3 Market—Rebuilding America's Infrastructure*, New York: American International Group, April 2013. As of June 8, 2018:
https://www.aig.com/content/dam/aig/america-canada/us/documents/insights/final-p3-aig -whitepaper-brochure.pdf

Michael, Gregory E., "Legal Issues Including Liability Associated with the Acquisition, Use, and Failure of GPS/GNSS," *Journal of Navigation*, Vol. 52, No. 2, 1999, pp. 246–251.

Microsemi, "Timing and Synchronization for LTE-TDD and LTE-Advanced Mobile Networks," Aliso Viejo, Calif.: Microsemi, White Paper, 2014. As of October 13, 2019:
https://www.microsemi.com/document-portal/doc_view/133615-timing-sync-for-lte-tdd-lte-a -mobile-networks

———, "SA.45s CSAC Options 001 and 003: Chip-Scale Atomic Clock," Aliso Viejo, Calif.: Microsemi, April 2018.

Microsoft, "Windows Server 2016 Accurate Time," webpage, December 21, 2016. As of October 13, 2019:
https://docs.microsoft.com/en-us/windows-server/identity/ad-ds/get-started/windows-time-service/ accurate-time

Milner, Greg, "What Would Happen if G.P.S. Failed?" *The New Yorker*, May 6, 2016. As of June 26, 2016:
https://www.newyorker.com/tech/elements/what-would-happen-if-gps-failed

Mitch, R. H., R. C. Dougherty, M. L. Psiaki, S. P. Powell, B. W. O'Hanlon, J. A. Bhatti, and T. E. Humphreys, "Signal Characteristics of Civil GPS Jammers," in *Proceedings of the 24th International Technical Meeting of the Satellite Division of the Institute of Navigation (ION GNSS 2011)*, Portland, Oreg., September 20–23, 2011.

Moore, Samuel K., "Superaccurate GPS Chips Coming to Smartphones in 2018," *IEEE Spectrum*, September 21, 2017. As of October 13, 2019:
https://spectrum.ieee.org/tech-talk/semiconductors/design/
superaccurate-gps-chips-coming-to-smartphones-in-2018

Morales, J. J., P. F. Roysdon, and Z. M. Kassas, "Signals of Opportunity Aided Inertial Navigation," presentation at the 29th International Technical Meeting of the Satellite Division of the Institute of Navigation (ION GNSS+ 2016), Portland, Oreg., September 12–16, 2016.

Moselle, Ben, ed., *2017 National Building Cost Manual*, 41st ed., Carlsbad, Calif.: Craftsman Book Company, 2017.

"N. Korean GPS Jamming Threatens Passenger Planes," *The Chosun Ilbo*, May 10, 2012. As of October 13, 2019:
http://english.chosun.com/site/data/html_dir/2012/05/10/2012051000917.html

Narins, Mitch, "The Global Loran/eLoran Infrastructure Evolution: A Robust and Resilient PNT Backup for GNSS," presentation at the 13th Meeting of the Space-Based Positioning, Navigation, and Timing National Advisory Board, Washington, D.C., June 3–4, 2014.

National 911 Program, *2017 National 911 Progress Report*, Washington, D.C.: 911.gov, November 2017. As of October 13, 2019:
https://www.911.gov/pdf/National-911-Program-Profile-Database-Progress-Report-2017.pdf

National Coordination Office for Space-Based Positioning, Navigation, and Timing, "U.S. Space-Based Positioning, Navigation, and Timing Policy: Fact Sheet," GPS.gov, December 15, 2004. As of December 11, 2018:
https://www.gps.gov/policy/docs/2004/

———, "Agriculture," GPS.gov, 2018. As of June 23, 2018:
https://www.gps.gov/applications/agriculture

National Consortium for the Study of Terrorism and Responses to Terrorism, Global Terrorism Database, College Park, Md.: University of Maryland, 2018. As of October 13, 2019:
https://www.start.umd.edu/gtd

National Emergency Number Association, "9-1-1 Statistics," webpage, 2018. As of October 13, 2019:
https://www.nena.org/page/911Statistics

National Institute of Standards and Technology, "Common View GPS Time Transfer," webpage, August 25, 2016. As of October 13, 2019:
https://www.nist.gov/pml/time-and-frequency-division/atomic-standards/common-view-gps
-time-transfer

———, "NIST Internet Time Service," webpage, January 3, 2018. As of October 13, 2019:
https://www.nist.gov/pml/time-and-frequency-division/services/internet-time-service-its

———, "Time and Frequency Services," webpage, May 2, 2019. As of October 13, 2019:
https://www.nist.gov/pml/time-and-frequency-division/time-services

National Marine Manufacturers Association, "U.S. Boating Industry Sales at 10-Year-High," Chicago, Ill.: NMMA, May 23, 2018. As of October 13, 2019:
https://www.nmma.org/press/article/21974

National Oceanic and Atmospheric Administration, *U.S. Billion-Dollar Weather and Climate Disasters*, Asheville, N.C.: National Centers for Environmental Information, 2017. As of October 13, 2019:
https://www.ncdc.noaa.gov/billions/

National Security Space Office, *National Positioning, Navigation, and Timing Architecture Study: Final Report*, Washington, D.C.: National Security Space Office, September 2008. As of December 19, 2018:
https://apps.dtic.mil/dtic/tr/fulltext/u2/a494613.pdf

National Transportation Safety Board, *Derailment of Amtrak Train NO. 2 on the CSXT Big Bayou Canot Bridge*, Washington, D.C.: NTSB, RAR-94-01, September 19, 1994. As of October 13, 2019:
https://www.ntsb.gov/investigations/accidentreports/pages/RAR9401.aspx

National Transportation Safety Board, *Collision of Metrolink Train 111 with Union Pacific Train LOF65-12*, Washington, D.C.: NTSB, RAR-10-01, January 21, 2010. As of October 13, 2019:
https://www.ntsb.gov/investigations/accidentreports/pages/RAR1001.aspx

NENA—*See* National Emergency Number Association.

NextNav, "Precise Indoor Location Is Now Within First Responders' Reach," Sunnyvale, Calif.: NextNav, Overview Sheet, 2013. As of October 13, 2019:
http://www.nextnav.com/sites/default/files/CSRIC%20Results%20One-Sheet.pdf

———, "Terrestrial Beacon System: 3D Indoor Geolocation," Sunnyvale, Calif.: NextNav, February 17, 2015. As of October 13, 2019:
https://pubs.naruc.org/pub.cfm?id=4AD5FBA2-2354-D714-51EB-8B5B29ECF6A4

Nilsen, Thomas, "Pilots Warned of Jamming in Finnmark," *The Barents Observer*, November 2, 2018. As of October 13, 2019:
https://thebarentsobserver.com/en/security/2018/11/pilots-warned-jamming-finnmark

NIST—*See* National Institute of Standards and Technology.

NOAA—*See* National Oceanic and Atmospheric Administration.

North American Electric Reliability Corporation, "Extended Loss of GPS Impact on Reliability," Atlanta, Ga.: NERC, White Paper, 2017.

Novick, A. N., and M. A. Lombardi, "Practical Limitations of NTP Time Transfer," presentation at the 2015 Joint Conference of the IEEE International Frequency Control Symposium & European Frequency and Time Forum, Denver, Colo., April 2015.

"NPL Offers Precise Timing for Intercontinental Exchange Data Centre in Basildon, UK," *International Finance*, May 9, 2017. As of October 13, 2019:
https://internationalfinance.com/npl-offers-precise-timing-intercontinental-exchange-data-centre-basildon-uk/

NTSB—*See* National Transportation Safety Board.

Oak Ridge National Laboratory, "Requirements for a Secure and Resilient National Terrestrial Time Source," Oak Ridge, Tenn.: ORNL, Response to DOT-OST-2016-0226, 2016.

Ochieng, W. Y., K. Sauer, P. A. Cross, K. F. Sheridan, J. Iliffe, S. Lannelongue, N. Ammour, and K. Petit, "Potential Performance Levels of a Combined Galileo/GPS Navigation System," *Journal of Navigation*, Vol. 54, No. 2, 2001, pp. 185–197.

O'Connor, A., M. Gallaher, K. Clark-Sutton, D. Lapidus, Z. Oliver, T. Scott, D. Wood, M. Gonzalez, E. Brown, and J. Fletcher, *Economic Benefits of the Global Positioning System (GPS): Final Report*, Research Triangle Park, N.C.: RTI International, RTI Report No. 0215471, June 2019.

Odenwald, Sten, "The Day the Sun Brought Darkness," NASA, March 13, 2009. As of October 13, 2019:
https://www.nasa.gov/topics/earth/features/sun_darkness.html

Offermans, G., S. Bartlett, and C. Schue, "Proving a Resilient Timing and UTC Service Using eLoran in the United States," *Journal of the Institute of Navigation*, Vol. 64, 2017, pp. 339–349.

Office of Management and Budget, "Appendix C: Discount Rates for Cost-Effectiveness, Lease-Purchase, and Related Analyses," Washington, D.C.: OMB, OMB Circular No. A-94, November 2017. As of June 7, 2018:
https://www.whitehouse.gov/wp-content/uploads/2017/11/Appendix-C-revised.pdf

———, *North American Industry Classification System: United States, 2017*, Washington, D.C.: Executive Office of the President, Office of Management and Budget, 2018. As of October 13, 2019:
https://www.census.gov/eos/www/naics/2017NAICS/2017_NAICS_Manual.pdf

OMB—*See* Office of Management and Budget.

Orolia, "VersaPNT Assured PNT Solution," webpage, undated. As of May 2, 2019:
https://www.orolia.com/products-services/rugged-pnt-sources/versapnt

Osei-Kyei, Robert, and Albert P. C. Chan, "Review of Studies on the Critical Success Factors for Public-Private Partnership (PPP) Projects from 1990 to 2013," *International Journal of Project Management*, Vol. 33, 2015, pp. 1335–1346.

Oxera Consulting, *What is the Economic Impact of Geo Services?* Oxford, U.K.: Oxera Consulting, report prepared for Google, January 2013. As of October 13, 2019:
https://www.oxera.com/Oxera/media/Oxera/downloads/reports/What-is-the-economic-impact-of-Geo-services_1.pdf

Park, JiYoung, "The Economic Impacts of Dirty Bomb Attacks on the Los Angeles and Long Beach Ports: Applying the Supply-Driven NIEMO (National Interstate Economic Model)," *Journal of Homeland Security and Emergency Management*, Vol. 5, No. 1, May 2008.

Park, JiYoung, Peter Gordon, James Moore, and Harry W. Richardson, "The State-by-State Economic Impacts of the 2002 Shutdown of the Los Angeles–Long Beach Ports," *Growth and Change*, Vol. 39, No. 4, 2008.

Parker, George, Peggy Hollinger, and Alex Barker, "UK Explores Producing Own Satellite System After EU's Galileo Snub," *Financial Times*, April 24, 2018.

Parkinson, Bradford, James Doherty, John Darrah, Arnold Donahue, Leon Hirsch, Donald Jewell, William Klepczynski, Judah Levine, Kirk Lewis, Edwin Stear, Phillip Ward, and Pamela Rambow, *Independent Assessment Team (IAT) Summary of Initial Findings on eLoran*, Alexandria, Va.: Institute for Defense Analysis, January 2009.

Patel, Nitesh, *Consumer Attitudes towards the Emerging Location Opportunity*, Newton, Mass.: Strategy Analytics, April 2009. As of October 13, 2019:
https://www.nottingham.ac.uk/grace/documents/resources/marketreports/consumerattitudelocationopportunity08.pdf

Pew Research Center, "Mobile Fact Sheet: Mobile phone ownership over time," Washington, D.C.: Pew Research Center, Fact Sheet, June 12, 2019. As of October 13, 2019:
https://www.pewinternet.org/fact-sheet/mobile/

Pham, Nam Dinh, *The Economic Benefits of Commercial GPS Use in the U.S. and the Costs of Potential Disruption*, New York: NDP Consulting, June 2011. As of October 13, 2019: http://www.saveourgps.org/pdf/GPS-Report-June-22-2011.pdf

Phillips, Tony, "Getting Ready for the Next Big Solar Storm," NASA, June 22, 2011. As of October 13, 2019: https://www.nasa.gov/mission_pages/sunearth/news/next-solarstorm.html

Positioning, Navigation, and Timing (PNT) Service for National Critical Infrastructure Resiliency, 81 Fed. Reg. 86378, November 30, 2016.

Powers, E., and A. Colina, "Wide Area Wireless Network Synchronization Using Locata," Washington, D.C.: U.S. Naval Observatory, 2015a. As of October 13, 2019: http://www.locata.com/wp-content/uploads/2015/09/USNO-Wide-Area-Wireless-Network -Synchronization-Using-Locata-ION-Publication-9-20-2015.pdf

———, "Timing Accuracy Down to Picoseconds," *GPS World*, October 5, 2015b.

Public Law 112-96, Middle Class Tax Relief and Job Creation Act of 2012, Title VI—Public Safety Communications and Electromagnetic Spectrum Auctions, February 22, 2012.

Purton, Leon, Hussein Abbass, and Sameer Alam, "Identification of ADS-B System Vulnerabilities and Threats," in *ATRF 2010: 33rd Australasian Transport Research Forum 2010 Proceedings*, Canberra: ATRF, 2010.

Rao, M., and G. Falco, "GNSS Solutions: Code Tracking & Pseudoranges," *Inside GNSS*, January/February 2012. As of October 13, 2019: http://insidegnss.com/wp-content/uploads/2018/01/IGM_janfeb12-Solutions.pdf

Raquet, John F., "Navigation Using Pseudolites, Beacons, and Signals of Opportunity," Brussels: NATO, Science and Technology Organization, Report STO-EN-SET-197-08, October 15, 2013.

Reelektronika, "Loradd++/core: eLoran/Chayka receiver module," Reeuwijk, Netherlands: Reelktronika, Specifications notice, March 15, 2017. As of October 13, 2019: https://www.reelektronika.nl/manuals/Loradd++_flyer.pdf

Renfro, Brent A., Audric Terry, and Nicholas Boeker, *An Analysis of Global Positioning System (GPS) Standard Positioning System (SPS) Performance for 2016*, Austin, Tex.: The University of Texas at Austin Applied Research Laboratories Space and Geophysics Laboratory, TR-SGL-17-06, May 2017. As of October 13, 2019: https://www.gps.gov/systems/gps/performance/2016-GPS-SPS-performance-analysis.pdf

Resilient Navigation and Timing Foundation, *Prioritizing Dangers to the United States from Threats to GPS*, Alexandria, Va.: RNT Foundation, White Paper, November 30, 2016. As of October 13, 2019: https://rntfnd.org/wp-content/uploads/12-7-Prioritizing-Dangers-to-US-fm-Threats-to-GPS -RNTFoundation.pdf

Reuters, "Iridium Launches Timing, Location Service as GPS Back-up," *Technology News*, May 23, 2016. As of October 13, 2019: https://www.reuters.com/article/us-iridium-gps/iridium-launches-timing-location-service-as-gps -back-up-idUSKCN0YE1HZ

Riahi Manesh, Mohsen, and Naima Kaabouch, "Analysis of Vulnerabilities, Attacks, Countermeasures and Overall Risk of the Automatic Dependent Surveillance-Broadcast (ADS-B) System," *International Journal of Critical Infrastructure Protection*, Vol. 19, 2017, pp. 16–31.

Riley, P., "On the Probability of Occurrence of Extreme Space Weather Events," presentation to SSB Committee on Solar and Space Physics, San Diego, Calif., October 7, 2014. As of October 13, 2019: http://sites.nationalacademies.org/cs/groups/ssbsite/documents/webpage/ssb_153147.pdf

Rizos, C., N. Gambale, and B. Lilly, "Synchronized Ground Networks Usher in Next-Gen GNSS," *GPS World*, October 2013, pp. 36–42.

Roessler, A., "Pre-5G and 5G: Will the mmWave Link Work?" *Microwave Journal*, December 13, 2017.

Rolli, Joe, "GPS Interference Detection & Geolocation Technology," presentation at the 16th Meeting of the Space-Based Positioning, Navigation, and Timing National Advisory Board, Boulder, Colo., October 30–31, 2015. As of October 13, 2019:
https://www.gps.gov/governance/advisory/meetings/2015-10/rolli.pdf

Rose, Adam, Gbadebo Oladosu, and Shu-Yi Liao, "Business Interruption Impacts of a Terrorist Attack on the Electric Power System of Los Angeles: Customer Resilience to a Total Blackout," *Risk Analysis*, Vol. 27, No. 3, 2007, pp. 513–531.

Rose, Adam, and Dan Wei, "Estimating the Economic Consequences of a Port Shutdown: The Special Role of Resilience," *Economic Systems Research*, Vol. 25, No. 2, 2013.

Roskind, Frank D., *Positive Train Control Systems: Regulatory Impact Analysis*, Washington, D.C.: U.S. Department of Transportation, 2009.

Royal Academy of Engineering, *Extreme Space Weather: Impacts on Engineered Systems and Infrastructure*, London: Royal Academy of Engineering, 2013. As of October 13, 2019:
https://www.raeng.org.uk/publications/reports/space-weather-full-report

Sadlier, Greg, Rasmus Flytkjaer, Farooq Sabri, and Daniel Herr, *The Economic Impact on the UK of a Disruption to GNSS*, London: London Economics, June 2017. As of October 13, 2019:
https://www.gov.uk/government/uploads/system/uploads/attachment_data/file/619544/17.3254_Economic_impact_to_UK_of_a_disruption_to_GNSS_-_Full_Report.pdf

Safar, J. S., F. Vejrazka, and P. Williams, "Assessing the Limits of eLoran Positioning Accuracy," *International Journal on Marine Navigation and Safety of Sea Transportation*, Vol. 5, March 2011, pp. 93–101.

Sand, Stephan, Christian Mensing, and Armin Dammann, "Positioning in Wireless Communications Systems—Introduction and Overview," in *Proceedings of the 18th Wireless World Research Forum*, Chennai, India, November 5–7, 2007. As of October 13, 2019:
https://www.researchgate.net/publication/224985851_Positioning_in_Wireless_Communications_Systems_-_Introduction_and_Overview

Satelles, "Satelles Time and Location," Herndon, Va.: Satelles, White Paper, May 2016. As of October 13, 2019:
http://www.satellesinc.com/wp-content/uploads/2016/05/Satelles-White-Paper-Final.pdf

Saul, Jonathan, "Europe Calls Time on Ship Navigation Scheme Despite Risks at Sea," Reuters, February 9, 2016. As of October 13, 2019:
https://uk.reuters.com/article/uk-shipping-navigation-gps/europe-calls-time-on-ship-navigation-scheme-despite-risks-at-sea-idUKKCN0VI0YW

———, "Cyber Threats Prompt Return of Radio for Ship Navigation," Reuters, August 6, 2017. As of October 13, 2019:
https://uk.reuters.com/article/us-shipping-gps-cyber/cyber-threats-prompt-return-of-radio-for-ship-navigation-idUKKBN1AN0HT

Saw, John, "The TDD-LTE Advantage," Sprint Blog, February 24, 2014. As of October 13, 2019:
https://newsroom.sprint.com/the-tdd-lte-advantage-1.htm

Schimmelpfennig, David, *Farm Profits and Adoption of Precision Agriculture*, Washington, D.C.: U.S. Department of Agriculture, Economic Research Report No. 217, October 2016. As of October 13, 2019:
https://www.ers.usda.gov/webdocs/publications/80326/err-217.pdf?v=42661

Schue, C. "Indoor Enhanced Loran: Demonstrating Secure Accurate Time at the NYSE," presentation at UrsaNav briefing, New York, April 19, 2016. As of October 13, 2019:
http://www.ursanav.com/wp-content/uploads/NYSE_Seminar_UrsaNav_19APR2016.pdf

Scott, Logan, "The Portland Spoofing Incident: An Example of Misplaced Trust," presentation at the 20th Meeting of the Space-Based Positioning, Navigation, and Timing National Advisory Board, Redondo Beach, Calif., November 16, 2017. As of October 13, 2019:
https://www.gps.gov/governance/advisory/meetings/2017-11/scott2.pdf

SDR Touch, homepage, undated. As of October 13, 2019:
http://sdrtouch.com

"Seoul's Makeshift Answer to N. Korean Jamming Attacks," *The Chosun Ilbo*, September 23, 2011. As of October 13, 2019:
http://english.chosun.com/site/data/html_dir/2011/09/23/2011092300630.html

Serrano, J., M. Cattin, E. Gousiou, E. van der Bij, T. Wlostowoski, G. Daniluk, and M. Lipinski, "The White Rabbit Project," in *Proceedings of 2nd International Beam Instrumentation Conference*, Oxford, U.K., September 16–19, 2013. As of October 13, 2019:
http://accelconf.web.cern.ch/accelconf/ibic2013/papers/thbl2.pdf

Shamaei, K., J. Lhalife, and J. M. Kassas, "Performance Characterization of Positioning in LTE Systems," in *Proceedings of the 29th International Technical Meeting of the Satellite Division of the Institute of Navigation (ION GNSS+ 2016)*, Portland, Oreg., September 12–16, 2016.

Shkel, A. M., "Precision Navigation and Timing Enabled by Microtechnology: Are We There Yet?" in T. George, M. Saif Islam, and A. K. Dutta, eds., *Micro- and Nanotechnology Sensors, Systems, and Applications III*, Bellingham, Wash.: International Society for Optical Engineering, 2011.

Siceloff, Steven, "Shuttle Computers Navigate Record of Reliability," NASA, June 28, 2010. As of October 13, 2019:
https://www.nasa.gov/mission_pages/shuttle/flyout/flyfeature_shuttlecomputers.html

Skyhook Wireless, "Coverage Area," webpage, undated. As of October 13, 2019:
https://www.skyhookwireless.com/Coverage-Map

Skey, Kevin M., "Responsible Use of GPS for Critical Infrastructure," presentation at the Critical Infrastructure Protection and Resilience North America Conference, Kennedy Space Center, Fla., December 6, 2017.

Small, D., and Locata Corp., "Method and device for chronologically synchronizing a location network," U.S. Patent 7,616,682, filed November 2, 2001, and issued November 10, 2009.

Smith, Benjamin, "The Economic Benefits of Precision Agriculture," presentation at the Tenth Meeting of the Space-Based Positioning Navigation & Timing National Advisory Board, August 14–15, 2012. As of October 13, 2019:
https://www.gps.gov/governance/advisory/meetings/2012-08/smith2.pdf

Son, Pyo-Woong, Sang Hyun Park, Kiyeol Seo, Younghoon Han, Jiwon Seo, "Development of the Korean eLoran Testbed and Analysis of Its Expected Positioning Accuracy," presentation at the 19th International Association of Marine Aids to Navigation and Lighthouse Authorities Conference, Incheon, South Korea, May 27–June 2, 2018. As of October 13, 2019:
https://rntfnd.org/wp-content/uploads/Korea-eLoran-2018.IALA_.pdf

Spectracom, "Frequently Asked Questions About STL," webpage, April 18, 2018. As of October 13, 2019:
https://spectracom.com/products-services/precision-timing/stl-faq

START—*See* National Consortium for the Study of Terrorism and Responses to Terrorism.

Statistics Korea, "Results of the Farm and Fishery Household Economy Survey in 2016," press release, Daejeon, April 24, 2017. As of October 13, 2019:
http://kostat.go.kr/portal/eng/pressReleases/2/8/index.board

Stevens, Laura, and Paul Ziobro, "Ports Gridlock Reshapes the Supply Chain," *The Wall Street Journal*, March 5, 2015. As of October 13, 2019:
https://www.wsj.com/articles/ports-gridlock-reshapes-the-supply-chain-1425567704

Streitwieser, Mary L., *Measuring the Nation's Economy: An Industry Perspective—A Primer on BEA's Industry Accounts*, Washington, D.C.: Bureau of Economic Analysis, 2011.

Tariq, Zain Bin, Dost Muhammad Cheema, Muhammad Zahir Kamran, and Ijaz Haider Naqvi, "Non-GPS Positioning Systems: A Survey," *ACM Computing Surveys*, Vol. 50, No. 4, August 2017.

Taylor, Heather, "Cost of Constructing a Home," National Association of Home Builders Economics and Housing Policy Group, November 2, 2015. As of October 13, 2019:
https://www.nahbclassic.org/generic.aspx?genericContentID=248306

Technical Strategy Leadership Group, *The Future Railway: The Industry's Rail Technical Strategy 2012*, United Kingdom, December 14, 2012. As of October 13, 2019:
https://www.ucl.ac.uk/resilience-research/research/railwaysignal/UCLWorkshop19Jun2013RSSB

Tefft, B. C., "American Driving Survey, 2015–2016," Washington, D.C.: AAA Foundation for Traffic Safety, January 2018. As of October 13, 2019:
https://aaafoundation.org/american-driving-survey-2015-2016/

Thompson, Wayne, Wei Chen, and Christopher Lawson, "Positioning and Navigation Market Assessment Across 15 Critical Infrastructure Sectors," Aerospace Corporation, January 31, 2018.

Tralli, David M., John G. Paul, Wayne K. Thompson, William W. Chen, Christopher M. Lawson, Victor S. Lin, Hemanshu Patel, Paul D. Massatt, and Andrew J. Binder, *Positioning and Navigation Critical Infrastructure Market Assessment and User Needs Framework and Methodology: Towards Developing a Formal Set of Structured, Validated Requirements and Technical Specifications for a Complementary or Backup System to the U.S. Global Positioning System*, El Segundo, Calif.: Aerospace Corporation, April 20, 2018a.

———, *Annex: Transportation Sector, Positioning and Navigation Critical Infrastructure Market Assessment and User Needs Framework and Methodology: Towards Developing a Formal Set of Structured, Validated Requirements and Technical Specifications for a Complementary or Backup System to the U.S. Global Positioning System*, El Segundo, Calif.: Aerospace Corporation, November 30, 2018b.

Tsurutani, B. T., W. D. Gonzalez, G. S. Lakhina, and S. Alex, "The Extreme Magnetic Storm of 1–2 September 1859," *Journal of Geophysical Research*, Vol. 108, No. A7, 2003.

Tuttle, Brad, "$47 a Month? Why You're Probably Paying Double the 'Average' Cell Phone Bill," *Time*, October 18, 2012. As of June 21, 2018:
http://business.time.com/2012/10/18/47-a-month-why-youre-probably-paying-double-the-average-cell-phone-bill/

U.K. Government Office for Science, *Satellite-Derived Time and Position: A Study of Critical Dependencies*, London: Government Office for Science, January 30, 2018. As of October 13, 2019: https://assets.publishing.service.gov.uk/government/uploads/system/uploads/attachment_data/file/676675/satellite-derived-time-and-position-blackett-review.pdf

U.K. National Audit Office, *PFI and PF2*, London: National Audit Office, Report by the Comptroller and Auditor General, January 17, 2018. As of June 7, 2018: https://www.nao.org.uk/wp-content/uploads/2018/01/PFI-and-PF2.pdf

Ulfarsson, Gudmundur Freyr, and Elizabeth A. Unger, "Impacts and Responses of Icelandic Aviation to the 2010 Eyjafjallajökull Volcanic Eruptions: Case Study," *Transportation Research Record*, Vol. 2214, No. 1, 2011.

Unified Facilities Criteria, *Programming Cost Estimates for Military Construction*, Washington, D.C.: US Army Corps of Engineers, Naval Facilities Engineering Command, and Air Force Civil Engineer Support Agency, UFC 3-730-01, March 1, 2017. As of October 13, 2019: https://www.wbdg.org/FFC/DOD/UFC/ufc_3_730_01_2011_c1.pdf

Union Pacific, "Positive Train Control," webpage, undated. As of October 13, 2019: https://www.up.com/media/media_kit/ptc/about-ptc/

UrsaNav Inc., and Harris Corporation, "CRADA2 Test Results," presentation, January 2017. As of October 13, 2019: http://www.ursanav.com/wp-content/uploads/CRADA2-eLoran-Test-Results-January-2017.pdf

U.S. Army Corp of Engineers, Institute for Water Resources, *Deep Water Ports and Harbors: Value to the Nation*, Washington, D.C.: U.S. Army Corps of Engineers, 2009.

———, *U.S. Port and Inland Waterways Modernization: Preparing for Post-Panamax Vessels*, Washington, D.C.: U.S. Army Corp of Engineers, June 20, 2012. As of October 13, 2019: https://www.iwr.usace.army.mil/Portals/70/docs/portswaterways/rpt/June_20_U.S._Port_and_Inland_Waterways_Preparing_for_Post_Panamax_Vessels.pdf

U.S. Census Bureau, "Metropolitan and Micropolitan—Population Density by Census Tract: 2010," *Census.gov*, undated. As of August 26, 2019: https://www.census.gov/data-tools/demo/metro-micro/thematic_maps.html

———, "Finance and Insurance: Subject Series–Estab & Firm Size: Summary Statistics by Concentration of Largest Firms for the U.S.: 2012," in *2012 Economic Census*, Washington, D.C.: U.S. Census Bureau, 2012. As of October 13, 2019: https://www.census.gov/data/tables/2012/econ/census/fincance-insurance.html

———, "2012 SUSB Annual Datasets by Establishment Industry," *Census.gov*, October 3, 2016. As of October 13, 2019: https://www.census.gov/data/datasets/2012/econ/susb/2012-susb.html

———, "Annual Estimates of the Resident Population: April 1, 2010 to July 1, 2017—United States—Metropolitan Statistical Area; and for Puerto Rico: 2017 Estimates," American FactFinder, 2018. As of June 21, 2018: https://factfinder.census.gov/faces/tableservices/jsf/pages/productview.xhtml?src=bkmk

U.S. Coast Guard, *Loran-C User Handbook*, Washington, D.C.: U.S. Department of Transportation, Commandant Publication P16562.5, 1992.

U.S. Coast Guard Navigation Center, "Loran-C Status Information," webpage, March 17, 2011. As of October 13, 2019: https://www.navcen.uscg.gov/?pageName=loranStatus,

USDA—*See* U.S. Department of Agriculture.

U.S. Department of Agriculture, "USDA's National Agricultural Statistics Service, Kansas Field Office: State and County Data," webpage, 2016. As of October 13, 2019:
https://www.nass.usda.gov/Statistics_by_State/Kansas/index.php

———, "USDA's National Agricultural Statistics Service, Kansas Field Office: County Estimates—Crops—Winter Wheat," webpage, 2017. As of October 13, 2019:
https://www.nass.usda.gov/Statistics_by_State/Kansas/Publications/County_Estimates/index.php

U.S. Department of Agriculture, National Agricultural Statistics Service, *Field Crops: Usual Planting and Harvesting Dates*, Washington, D.C.: USDA, Agricultural Handbook No. 628, October 2010. As of October 13, 2019:
https://downloads.usda.library.cornell.edu/usda-esmis/files/vm40xr56k/dv13zw65p/w9505297d/planting-10-29-2010.pdf

———, *Farms and Farmland, 2012 Census of Agriculture Highlights*, Washington, D.C.: USDA, ACH12-13, September 2014. As of October 13, 2019:
https://www.nass.usda.gov/Publications/Highlights/2014/Highlights_Farms_and_Farmland.pdf

U.S. Department of Commerce, "Common View GPS Time Transfer, National Institute of Standards and Technology (NIST)," webpage, August 25, 2016. As of October 13, 2019:
https://www.nist.gov/pml/time-and-frequency-division/atomic-standards/common-view-gps-time-transfer

U.S. Department of Defense, *Global Positioning System Standard Positioning Service Performance Standard*, 4th ed., Washington, D.C.: U.S. Department of Defense, September 2008. As of October 13, 2019:
https://www.gps.gov/technical/ps/2008-SPS-performance-standard.pdf

———, *Technology Readiness Assessment (TRA) Guidance*, Washington, D.C.: Assistant Secretary of Defense for Research and Engineering, April 2011. As of October 13, 2019:
https://apps.dtic.mil/dtic/tr/fulltext/u2/a554900.pdf

U.S. Department of Defense, U.S. Department of Homeland Security, and U.S. Department of Transportation, *2017 Federal Radionavigation Plan*, Springfield, Va.: National Technical Information Service, DOT-VNTSC-OST-R-15-01, 2017. As of October 13, 2019:
https://www.navcen.uscg.gov/pdf/FederalRadioNavigationPlan2017.pdf

U.S. Department of Homeland Security, *National Risk Estimate: Risks to U.S. Critical Infrastructure from Global Positioning System Disruptions*, Washington, D.C.: National Executive Committee for Space-Based Positioning, Navigation, and Timing, 2011.

———, *National Risk Estimate: Risks to U.S. Critical Infrastructure from Global Positioning System Disruptions*, Washington, D.C.: National Executive Committee for Space-Based Positioning, Navigation, and Timing, November 2012.

———, *Improving the Operation and Development of Global Positioning System (GPS) Equipment Used by Critical Infrastructure*, Washington, D.C.: National Cybersecurity & Communications Integration Center and National Coordinating Center for Communications, January 2017. As of October 13, 2019:
https://ics-cert.us-cert.gov/sites/default/files/documents/Improving_the_Operation_and_Development_of_Global_Positioning_System_%28GPS%29_Equipment_Used_by_Critical_Infrastructure_S508C.pdf

U.S. Department of Homeland Security, Science and Technology Directorate, "GPS Vulnerabilities for Critical Infrastructure," Washington, D.C.: U.S. Department of Homeland Security, Fact Sheet, April 13, 2016.

U.S. Department of Transportation, *Global Positioning System (GPS) Civil Monitoring Performance Specification*, El Segundo, Calif.: Global Positioning Systems Wing/GPC, DOT-VNTSC-FAA-09-08, April 30, 2009. As of October 13, 2019:
https://www.gps.gov/technical/ps/2009-civil-monitoring-performance-specification.pdf

———, *Revised Departmental Guidance 2016: Treatment of the Value of Preventing Fatalities and Injuries in Preparing Economic Analyses*, Washington, D.C.: U.S. Department of Transportation, August 8, 2016. As of October 13, 2019:
https://www.transportation.gov/office-policy/transportation-policy/revised-departmental-guidance
-on-valuation-of-a-statistical-life-in-economic-analysis

———, "Hazmat Intelligence Portal," website, 2018a. As of October 13, 2019:
https://hip.phmsa.dot.gov/

———, *Global Positioning System (GPS) Adjacent Band Compatibility Assessment: Final Report*, Washington, D.C.: U.S. Department of Transportation, April 2018b. As of October 13, 2019:
https://www.transportation.gov/sites/dot.gov/files/docs/subdoc/186/dot-gps-adjacent-band-final
-reportapril2018.pdf

U.S. Department of Transportation, Bureau of Transportation Statistics, *Freight Facts and Figures 2017*, Washington, D.C.: U.S. Department of Transportation, October 13, 2017.

U.S. Department of Transportation, Federal Railroad Administration, Office of Safety Analysis, "Train Fatalities, Injuries and Accidents by Type of Accident(a)" (table), webpage, April 2018. As of October 13, 2019:
https://www.bts.gov/content/train-fatalities-injuries-and-accidents-type-accidenta

U.S. Department of the Treasury, "10-Year High Quality Market (HQM) Corporate Bond Spot Rate (HQMCB10YR)" (figure), FRED Economic Data, Federal Reserve Bank of St. Louis, May 15, 2018. As of June 7, 2018:
https://fred.stlouisfed.org/series/HQMCB10YR

USGS—*See* U.S. Geological Survey.

U.S. Geological Survey, "Lidar Point Cloud—USGS National Map 3DEP Downloadable Data Collection," Washington, D.C.: U.S. Geological Survey, 2017.

———, "USGS EROS Archive—Digital Elevation—Shuttle Radar Topography Mission (SRTM) 1 Arc-Second Global," Washington, D.C.: U.S. Geological Survey, 2018.

U.S. Government Accountability Office, *National Risk Estimate: Risks to U.S. Critical Infrastructure from Global Positioning System Disruptions of 2012*, Washington, D.C.: GAO, 2013a.

———, *Global Positioning System: A Comprehensive Assessment of Potential Options and Related Costs Is Needed*, Washington, D.C.: GAO, 2013b.

———, *GPS Disruptions: Efforts to Assess Risks to Critical Infrastructure and Coordinate Agency Actions Should Be Enhanced*, Washington, D.C.: GAO, GAO-14-15, November 2013c.

———, *Critical Infrastructure Protection: Electricity Suppliers Have Taken Actions to Address Electromagnetic Risks, and Additional Research Is Ongoing*, Washington, D.C.: GAO, 2018.

van Dongen, Teun, "The Lengths Terrorists Go To: Perpetrator Characteristics and the Complexity of Jihadist Terrorist Attacks in Europe, 2004–2011," *Behavioral Sciences of Terrorism and Political Aggression*, Vol. 6, No. 1, 2014, pp. 58–80.

Van Dyke, Karen, "U.S. Department of Transportation (DOT) Civil GPS/PNT Update," presentation at the 16th Meeting of the National Space-Based Positioning, Navigation, and Timing National Advisory Board, Boulder, Colo., October 30–31, 2015.

van Graas, Frank, and Subbu Meiyappan, "Terrestrial GPS Augmentation with a Metropolitan Beacon System," presentation at the 14th Meeting of the National Space-Based Positioning, Navigation, and Timing National Advisory Board, Washington, D.C., December 10, 2014, As of October 13, 2019:
https://www.gps.gov/governance/advisory/meetings/2014-12/vangraas-meiyappan.pdf

Van Sickle, Jan, "The Error Budget," University Park, Pa.: Pennsylvania State University, College of Earth and Mineral Sciences, GEOG 862: GPS and GNSS for Geospatial Professionals, 2018. As of October 13, 2019:
https://www.e-education.psu.edu/geog862/node/1712

Viscusi, K., and Joseph Aldy, "The Value of a Statistical Life: A Critical Review of Market Estimates Throughout the World," *Journal of Risk and Uncertainty*, Vol. 27, No. 1, 2003, pp. 5–76.

Wallischeck, Eric, *GPS Dependencies in the Transportation Sector: An Inventory of Global Positioning System Dependencies in the Transportation Sector, Best Practices for Improved Robustness of GPS Devices, and Potential Alternative Solutions for Positioning, Navigation and Timing*, Cambridge, Mass.: U.S. Department of Transportation Office of the Assistant Secretary for Research and Technology John A. Volpe National Transportation Systems Center, DOT-VNTSC-NOAA-16-01, 2016.

Ward, Nick, Chris Hargreaves, Paul Williams, and Martin Bransby, "Can eLoran Deliver Resilient PNT?" in *Proceedings of the ION 2015 Pacific PNT Meeting*, Honolulu, Hawaii, April 20–23, 2015.

Weiss, Marc, Lee Cosart, James Hanssen, and Jian Yao, "Precision Time Transfer using IEEE 1588 over OTN through a Commercial Optical Telecommunications Network," presented at the 2016 IEEE International Symposium on Precision Clock Synchronization for Measurement, Control, and Communication, Stockholm, Sweden, September 4–9, 2016.

Wertz, James R., and Wiley J. Larson, *Space Mission Analysis and Design*, 3rd ed., Torrance, Calif.: Microcosm Press; Dordrecht, Netherlands: Kluwer Academic, 1999.

The White House, "National Space Policy of the United States of America," Washington, D.C., June 28, 2010. As of October 13, 2019:
https://www.nasa.gov/sites/default/files/national_space_policy_6-28-10.pdf

World Bank, "Mobile Cellular Subscriptions (per 100 people)" (table), webpage, 2018. As of June 21, 2018:
https://data.worldbank.org/indicator/IT.CEL.SETS.P2

Xu, R., W. Chen, Y. Xu, and S. Hi, "A New Indoor Positioning System Architecture Using GPS Signals," *Sensors*, Vol. 15, 2015, pp. 10074–10087.

Yamanouchi, Kelly, "Airport Outage: Delta CEO to Seek Repayment for Lost Revenue," *Atlanta Journal-Constitution*, December 19, 2017.

Yang, Haomiao, Rongshun Huang, Xiaofen Wang, Jiang Deng, and Ruidong Chen, "EBAA: An Efficient Broadcast Authentication Scheme for ADS-B Communication Based on IBS-MR," *Chinese Journal of Aeronautics*, Vol. 27, No. 3, 2014, pp. 688–696.

Yonhap, "N. Korea's GPS Jamming Targeted at Aircraft Navigation Systems: Official," Yonhap News Agency, April 5, 2016a. As of October 13, 2019:
https://en.yna.co.kr/view/AEN20160405008851315

———, "N. Korea Halts GPS Jamming: Gov't," Yonhap News Agency, April 6, 2016b. As of October 13, 2019:
https://en.yna.co.kr/view/AEN20160406007800320

Young, Andrew, Christina Rogawski, and Stefaan Verhulst, *United States GPS System: Creating a Global Public Utility*, New York: GovLab; Washington, D.C.: Omidyar Network, January 2016. As of October 13, 2019:
http://odimpact.org/files/case-studies-gps.pdf

Zandbergen, Paul A., "Accuracy of iPhone Locations: A Comparison of Assisted GPS, WiFi and Cellular Positioning," *Transactions in GIS*, Vol. 13, No. s1, June 2009, pp. 5–25.

Zenko, Micah, "Dangerous Space Incidents," New York: Council on Foreign Relations, Contingency Planning Memorandum No. 21, April 16, 2014. As of October 13, 2019:
https://www.cfr.org/report/dangerous-space-incidents

Zhu, S., and D. M. Levinson, "Disruptions to Transportation Networks: A Review," in D. Levinson, H. Lui, and M. Bell, eds., *Network Reliability in Practice*, New York: Springer, 2012.

Interviews

Gates, Chris, general manager, NextNav, and Ganesh Pattabiraman, chief executive officer (CEO), NextNav, April 26, 2018. Interviewees discussed the MBS network cost overview and provided a proprietary brief with details related to the MBS system.

Kreger, Chris, representative, RF Specialties Group, April 12, 2018. Interviewee provided a rough-order-of-magnitude (ROM) cost estimate for a turnkey installation of a medium-wave (standard AM broadcast) transmitter, tower, control equipment, and building similar to an eLoran site as a cost cross-check to the UrsaNav proposal.

Lutter, Dan, general manager of North America, Seven Solutions S. L., May 31, 2018. Interviewee provided catalog pricing data for the various components necessary for a White Rabbit installation as well as a proprietary ROM cost estimate to install White Rabbit at all necessary eLoran locations.

Proctor, Andy, technical director, GNSS, U.K. Space Agency, May 16, 2018.

Schue, Charles, founder, president, and CEO of UrsaNav, Inc., May 10, 2018, and May 15, 2018.

Seo, Kiyeol, principal research engineer, Korea Research Institute of Ships and Ocean Engineering, May 28, 2018. In an email, the interviewee discussed the current and future status of eLoran in South Korea.